EYEWITNESS
TO THE GALAXY

EYEWITNESS TO THE GALAXY

Britain's contribution to research in space

PETER HAINING

W.H. ALLEN · LONDON
1985

Copyright © 1985 by Peter Haining

Typeset by Phoenix Photosetting, Chatham
Printed and bound in Spain by
Graficas Reunidas S.A., Madrid, Spain
for the Publishers, W.H. Allen & Co. PLC
44 Hill Street, London W1X 8LB

ISBN 0 491 03610 8

For

Barney Lodge

Who has been opening my
eyes to the wonders of
science for years

and

Bob Tanner

Who sent me on this
Mission of Discovery

ACKNOWLEDGEMENTS

I HAVE RECEIVED the help and co-operation of a great many people and organisations in Britain, Europe and America in the preparation of this book, and I should like to record my gratitude to the following. First and foremost, the National Aeronautics and Space Administration (NASA) in Washington, DC for their help with information on their own and other space programmes, and the generous provision of many of the illustrations; Ms Candi Chase of the Public Affairs Office was a knowledgeable and ever-helpful contact. The staff at the National Air and Space Museum in Washington were also most obliging with advice and information. For material on the Russian space programme I am grateful to the Novosti Press Agency and TASS, while the European Space Agency (ESA) and British Aerospace provided details of the European projects. For background information on the history of rocketry and space flights, I acknowledge with thanks the Library of Congress in Washington and the British Museum and British Newspaper Library in London. During the course of my research I consulted a great many newspapers, magazines, pamphlets and books, but these are too numerous to be mentioned here. And as far as individuals are concerned, I must also acknowledge Maxwell Woosnam, Harry Aldous, Arthur C. Clarke, Patrick Moore, Tom Smith, Peter Dacre, Robert L. Crippen, Joseph Allen, John W. Young, Derek Mullinger and Mike Ashley.

Opposite: *The British space team. From left: Major Richard Farrimond, Commander Peter Longhurst, Mr Christopher Holmes and Squadron Leader Nigel Wood, who is scheduled to become the first Briton in space; Commander Longhurst will be the second.* (ITN)

'An uninterrupted, navigable ocean that comes to the threshold of every man's door, ought not to be neglected as a source of human gratification and advantage.'

Sir George Cayley
The First Man to Fly a Heavier-than-air Machine

'It is difficult to say what is impossible, for the dream of yesterday is the hope of today and the reality of tomorrow.'

Robert H. Goddard
The Father of Modern Rocketry

'The completion of this flight opens new perspectives in the conquering of the Cosmos.'

Yuri Gagarin
The First Man in Space

CONTENTS

The Space Shuttle Discovery *taking off on its maiden flight from the Kennedy Space Center, Florida, 30 August 1984.*

INTRODUCTION

THEY DO MORE than dream of tomorrow in Washington DC – they turn those dreams into reality. For this beautiful, spacious city on the banks of the Potomac River is not just America's capital and the seat of government: it is also the headquarters of the National Aeronautics and Space Administration – NASA for short – which has already put men in space, and several on the Moon (12 to be precise), probed the planets of our solar system, and now plans to go further – much further. With a new telescope to look into the enormity of deep space, with space stations, bases on the Moon and even space colonies where men can live and work and prepare for the next giant leap across the frontier of the unknown on to the worlds of Mars and Jupiter.

It is due very much to the work of NASA that a Briton is at last going to join all those Americans (and Russians of course) who have travelled in space: an important moment in our history, to be sure, and one which this book sets out to commemorate. But as I found out during my visit to Washington, this particular mission is just one of many being planned at NASA's glass mountain of a headquarters on Maryland Avenue, where people talk of the solar system as if it were next door and the far reaches of space as if they were just across the road. Indeed, the visitor soon finds out that it is not only immense distances that NASA is conquering, but even time itself, for new ideas spring up here with breath-taking speed and become reality almost as quickly. Before reviewing the enormous amount that has already been achieved, however, it is worth looking at what lies ahead, hoping, of course, that these plans will not have become fact before this book reaches you!

EYEWITNESS TO THE GALAXY

The purpose of all these future projects can be put quite simply: to enable mankind to *see* further into the galaxy around him; *learn* more of the boundless possibilities of space and its worlds; and *travel* further than our dreams have ever dared allow us think was possible.

At the very heart of much of this planning lies the remarkable Space Shuttle, the reuseable orbital transporter which can leave and return to the Earth at will, and has quite rightly been called the 'first true space ship'. So far, it has been used to launch satellites, carry out observation of the Earth, and conduct a variety of experiments. Next it is to be used as the means of transporting the various components which will be erected in space to form, firstly, small space stations, and later whole space colonies. It will be vital, too, in missions to other planets (though not undertaking the actual journeys). Because of the success of this machine and its comparative cheapness to operate as compared to conventional rockets, NASA is already thinking in terms of about 500 flights for the existing four craft in the next dozen years.

The go-ahead for NASA's next great project, a space station, was given in January 1984 by President Reagan in his directive to the Agency to 'develop a permanently manned space station and to do it within a decade'. An instruction such as this – demanding the kind of finances that will probably call for a total investment of at least $8 billion by the time the station is ready for launching in the early 1990s – shows the same kind of bold and imaginative planning that led to the first man stepping on the Moon in 1969: just as the then President, John F. Kennedy, had directed a decade before. A permanent presence in space is felt to be an essential stepping-stone to the future and to all manner of experiments and projects of benefit to mankind.

Some NASA officials I spoke to believe it may be possible to have 'space factories' in operation even before the space station as a whole is launched. These units would provide the facilities for the commercial production of certain critical materials (medical drugs, for example) in conditions of low gravity and extreme purity not obtainable on Earth, as well as servicing the many satellites in orbit. In time, the units would be incorporated into the main space station.

'There is no question in my mind that within five years factories will be in orbit and will be served,' says J. Michael Smith, a NASA Senior Marketing

Development Specialist. 'The Shuttle has made this whole new industrial revolution possible by showing what can be done in a week or two of flight. The next step is to take things up there and leave them for a while and then go back and get them.'

According to the NASA plans I was shown, the space station concept envisages a facility in a low Earth orbit at an inclination of 28.5 degrees to the equator with provisions for a crew of six to eight people. In addition to living quarters the station will provide utilities (electrical power, thermal control, attitude control and data processing), work space, and a docking hub to allow entry and exit to the Space Shuttle. The Shuttle will enable the station crew to be rotated at three-to-six month intervals.

'The work of the space station will be conducted both in attached pressurised modules and on unpressurised free-flying platforms,' says the latest NASA report. 'The modules will be able to support scientific research and technology development requiring crew interaction. The unmanned platforms will be able to provide changeable payload accommodations for activities requiring minimum disturbance and protection from contamination. Our plans call for an initial space station to be operational early in the 1990s. It will be capable of growth both in size and capability and is intended to operate for several decades, well into the twenty-first century.'

A further benefit of the space station – and one seen as extremely important in many quarters – is that it will provide the facility for astronauts to remain in space for long periods and thereby gain the necessary experience for future projects such as lunar bases and manned missions to far-distant Mars and Jupiter. It will also be able to launch deep space probes to points of the universe currently beyond our knowledge.

Although there are scientific opponents to this plan for what has rather caustically been called a 'Cosmic Motel' – on the grounds that all the work could be done by robots – I personally believe it is an important step towards fulfilling the promise of space, and I am glad to be able to record that Britain is one of several nations that have indicated their interest in participating through the multi-national European Space Agency (who, in turn, hope to have their own space station one day).

Based on my discussions, I believe that the final space station will be rather like a scaled-down version of NASA 'Skylab' which has already been tried

Above: *Artist's impression of the exterior of a space colony for some 10,000 people, members of the workforce of a space manufacturing complex. The space station is built in the form of a rotating wheel.*

Top right: *An artist's conception of activity at a possible manned space station in the Earth's orbit.*

Bottom right: 'Columbia *at Booster Separation', an oil painting by Bob McCall.*

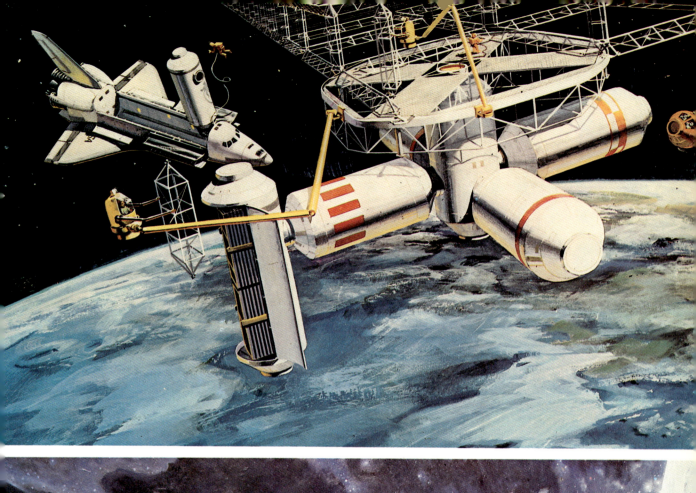

and tested. And knowing something of the Americans' love of timing, I would hazard a guess it will be launched to coincide with the 500th Anniversary of Columbus' discovery of America in 1492. It would surely be a most appropriate moment for the US to launch its own 'New World'!

I am encouraged in this view by the project's code name 'Columbus' and a remark President Reagan is said to have made when some of his aides questioned his decision to give the go-ahead for the space station. 'Just imagine,' he is reported to have replied, 'What would have happened if guys like you had interfered with Columbus asking Queen Isabella and King Ferdinand to help him go to the new world!'

These stations will, as I have indicated, lead in turn to the establishment of the enormous and complex space colonies which, as the imaginative and technically feasible illustrations in this book show, will one day provide complete new environments in space in which men and women will be able to live as naturally as they do on Earth.

Long before these miniature worlds in space are functioning, however, we shall certainly have bases on the Moon, as George Keyworth, President Reagan's science adviser has indicated. 'The space station is just a doorway to the Moon,' he says. 'I feel that the plusses are piling up pretty high on setting up a manned lunar station as the next major step.'

NASA scientists have, in fact, already begun to work on this project, and are convinced it would be quite possible to have an operational lunar base – ideally sited near one of the poles – by the year 2007. These experts say that the 'colonists' – about 6 to 8 people per base – would live and work in modules transported from Earth. They would be protected against cosmic rays by a covering of lunar dust which tests of the samples brought back by earlier missions have shown to consist of metals, silicons and oxygen. This dust could also be processed to provide raw materials for the inhabitants' needs. Fresh food, as such, could be supplied by pressurised 'greenhouses'.

An essential part of this plan is the development of Orbital Transfer Vessels (OTVs) which will be able to transfer to the Moon personnel and cargoes brought up from the Earth by the Shuttles. These vehicles, which NASA hopes to have developed by 1996, will be reuseable, able to be refueled in space, and should require less complicated manoeuvring when taking off and landing because of the Moon's airless surface. It has been estimated, in fact, that the energy cost of bringing materials from the Moon into space will be less than one-twentieth that of bringing them from Earth!

Using such a lunar base, say the NASA scientists, it will be possible to support large-scale developments in space such as the space colonies, solar power satellites, and, of course, spacecraft setting out to explore the universe.

However, an even more immediate project is the launching of the new NASA space telescope, named after the American astronomer, Edwin Hubble, which has been described as 'the greatest advance in astronomy since Galileo first pointed a telescope skyward 375 years ago.' This delicate optical assembly with its 94-inch diameter mirror (said to be the most perfect of its size) is to be put into an orbit 300 miles above the Earth's atmosphere by the newest of the Shuttles, *Atlantis*, in August 1986. This mission – code numbered 61-J – will realise a dream astronomers have had for twenty years: of placing an optical telescope in orbit free of the distortions caused to images by disturbances in our atmosphere.

With a lens aperture of four feet, the telescope will be able to observe small features on, say, the Sun, with *ten times* the magnification of ground-based instruments, and in all wave lengths from ultraviolet to infrared. It will be able to enlarge the amount of the universe currently observable by optical telescopes by an astonishing 350 times! This means astronomers will see across 14 billion light years to the edge of the universe, observing celestial objects 50 times fainter than is currently possible!

Once the 11-ton instrument has been launched from the cargo bay of the Shuttle, it is estimated that it will have a life in space of 15 years, although it will be possible for it to be serviced by astronauts. Controlled from Earth, its mission will be to add to our understanding of the age and size of the universe, the solar system, and the origins and evolution of galaxies. The Hubble telescope will surely give even greater range to our eyewitness view of space already being provided by the cameras of the far-ranging space probes.

There are also, I learn, plans to send two spacecraft to the very heart of our universe, the Sun, some time in the eighties. The Solar Polar Mission (ISPM) will look at the north and south poles of the Sun which we can only at present see from a very sharp angle. These vehicles will need to become the first

Above: *The establishment of a permanently manned space station will enable manned missions to be sent to Mars and Jupiter, following the unmanned* Voyager 1 *which took this picture of Jupiter in 1979.*

Top left: *Artist's impression of* Voyager 1 *as it passes through the strange phenomenon called the 'flux tube' of Io, one of the largest of Jupiter's satellites. The 'flux tube' is a region of magnetic and plasma interaction between the satellite and Jupiter.*

Bottom left: *Artist's impression of the* Galileo *spacecraft following its deployment from the Space Shuttle cargo bay.*

spacecraft to fly out of the eliptic – the plane in which the Earth and the other planets and moons in the solar system revolve around the Sun – but once having achieved this after a 16-month flight, they will then separate to pass 'up' and 'down' over the flaming body, completing their entire circumnavigations in six months.

As a result of this epic mission, the two craft will be able to relay from their unique viewpoints a vast amount of new data and pictures about this focal point of the life of the solar system. (NASA also has a project for a Solar Probe to penetrate the cornea of the Sun and from the searingly hot distance of 2.6 million miles send back information on its structure and composition.

The heat the craft will have to withstand to function at this distance has been estimated to be equal to that of about 2,500 Suns on our planetary surface!)

And still the ideas go on. As I mentioned earlier, schedules are being drawn up for manned landings on what appear to be the most hospitable planets in the solar system, Mars and Jupiter. NASA is also advancing its study of Black Holes and Pulsars, as well as investigating the first evidence of new solar systems hundreds of light years away.

Arguably, though, the most exciting scheme of all is Project SETI, the Search for Extraterrestrial Intelligence, which is bringing together the resources of the world's observatories for a systematic combining of space in the hope of answering that most intriguing of questions: 'Are We Alone?' Geared to reach the 'listening' stage in 1988 when it is hoped to pick up radio signals from somewhere in the cosmos, SETI will run for at least five years and possibly to the end of the century if any kind of positive response is received. Head of the organising committee, Harlan Smith, explains: 'There are few questions more important than whether the human race is alone in the universe. But the answer is going to be incredibly difficult to come by. It's worth a modest investment every year for the foreseeable future by techniques that will doubtless improve as time goes by. By great good fortune, we might succeed in learning something in the next few decades.' (NASA's 'modest' investment, incidentally, is $1.5 million per year.)

Another of the leading figures in the search, Britain's Professor Archibald Roy of Glasgow University is even more optimistic about the chances, although he accepts the possible 'targets' exceed 200 billion stars. 'I am sure civilisations are ten a penny around the cosmos,' he says. 'Indeed, in the vast cathedral of this universe, it is surely very peculiar to argue that we are its heroes. We are much more likely to be just bit players.'

And he adds convincingly, 'Within two or three years the hunt should be on in earnest. Then we can expect to make contact very quickly, probably within a decade.'

As part of Project SETI, the European Space Agency is also launching a highly sensitive space probe, *Hipparchus*, in 1988, which will be able to survey vast areas of the galaxy. Some of NASA's probes will augment the effort, too.

So you can see that planning and discovery go successfully hand-in-hand at

NASA, the place where dreams – even the most amazing and seemingly impossible – can come true. It is an exciting thought to ponder as the first Britons prepare to go into space, that we are all going to be a part of this adventure. And, as I hope to show in this book, the British have, through their contributions to the conquest of space over the years, *earned* a place, just like the Americans and the Russians.

NASA Headquarters,
Washington, DC.
December 1984.

The view from the unmanned Viking 2 *spaceship which landed on Mars in September 1976.*

THE STEVEDORES
OF SPACE

IT IS A BRIGHT, clear day in downtown Houston, the sky above is a vivid blue canopy that seems to sweep upwards forever – or at least to the very edge of space, which just happens to be the abiding preoccupation of this part of the state of Texas. For here the Old West has become the New West, and the frontier spirit is once more urging mankind to new and even greater achievements, day by day drawing in the vastness of our solar system. Why, even at night the Moon seems closer here than anywhere else on Earth.

The place in question is the Lyndon B. Johnson Space Center, a series of highly-modern white buildings set in 1,600 acres of flat, lush grasslands adjacent to a sparkling stretch of water known most appropriately as Clear Lake. Directly alongside is NASA Highway 1 which runs some twenty miles southeast to Houston itself, spreadeagled in an autumn haze. Although there is a town called Webster just two miles to the east and the Ellington Air Force Base seven miles away to the north, the Space Center seems to dominate the landscape and emphasise at once the claim that this is the country's most important location for the remarkable Space Shuttle.

And well-founded the claim soon proves to be. For here everything is organised from the moment a man or woman is chosen to fly in one of NASA's four Shuttles, the spacecraft which has captured the imagination of the world. Right through training and then the actual space mission itself –

Opposite: *As dawn breaks over the Kennedy Space Center, Florida, the Space Shuttle, complete with Mobile Launching Platform, is towed from the Vehicle Assembly Building to Launch Pad 39A.*

27

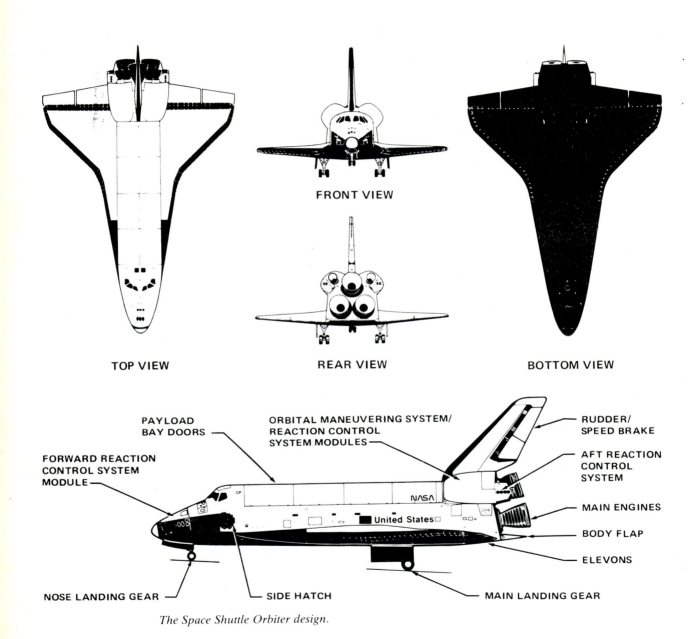

FRONT VIEW

TOP VIEW

REAR VIEW

BOTTOM VIEW

PAYLOAD
BAY DOORS

ORBITAL MANEUVERING SYSTEM/
REACTION CONTROL
SYSTEM MODULES

RUDDER/
SPEED BRAKE

FORWARD REACTION
CONTROL SYSTEM
MODULE

AFT REACTION
CONTROL
SYSTEM

MAIN ENGINES

NASA

United States

BODY FLAP

ELEVONS

NOSE LANDING GEAR

SIDE HATCH

MAIN LANDING GEAR

The Space Shuttle Orbiter design.

THE STEVEDORES OF SPACE

Houston is in charge. Here the embryo astronaut takes the first steps which will ultimately make him or her an eyewitness to the galaxy far beyond the confines of this Earth. It is in this 'palace of astronautical endeavour', as it has been called, that four of my fellow countrymen have trained for the distinction of becoming 'The First Briton In Space'.

Remarkable as it may at first seem, the Space Center was actually established a quarter of a century ago in September 1961 when it was set up as NASA's primary centre for the design, development and manufacture of manned spacecraft. Indeed, there are ever-present reminders of past triumphs all about the complex – space vehicles from virtually every manned programme including *Mercury, Gemini* and, of course, *Apollo*.

Apart from the training of Shuttle flight crews and the control of their missions, JSC also oversees many of the medical, scientific and engineering experiments which are carried out in space. It also directs operations at the White Sands Test Facility at Las Cruces in New Mexico which tests the Shuttle propulsion system, power system and materials; and most importantly of all directs the launching of the Shuttles at the John F. Kennedy Space Center close to Cape Canaveral on the east coast of Florida.

These launchings are in the vastly experienced hands of Mission Control Center situated at the very heart of the complex. Every manned mission starting with *Gemini IV* through *Apollo* and the *Skylab* series and on to the current Space Shuttle missions has been guided by this scientific wonderworld.

Of more immediate importance to new astronauts, however, is the Training Center, a huge hangar which contains among other things a mock-up of the Space Shuttle, and a totally authentic flight deck which can simulate an entire mission from launch to touch-down, complete with complex orbital manoeuvres and – if need be – emergencies. Here, too, they learn about their flight duties as well as more mundane things such as eating in a weightless environment, washing, shaving and going to the toilet. They even have to get used to sleeping upright in a restraining holster – that is if anything can be called *upright* in the topsy-turvey world of space travel!

It has been said by more than one astronaut that flying in space is rather like parachute jumping – you *have* to do it right the first time! So for this reason it is well worth taking a more detailed look at the training facilities.

Top left: *Johnson Space Center, Houston, Texas.*

Above: *The various stages of the Shuttle flight pattern, from take-off to landing.*

Below: *Mission Control Center, at the heart of the Johnson Space Center in Houston, Texas.*

Their objective is, quite simply, to put the crew of a Shuttle mission through a variety of simulators until their flight plan becomes second nature to them.

Firstly, there is the *One-G Trainer*, a full-scale flight deck, a mid-deck and mid-body (complete with the bay section for carrying the cargo or 'payload' as it is called). This is used for training the astronauts in their living conditions, in the methods of entry and exit from the Shuttle, television operations, waste management, stowage, general maintenance and what is delightfully termed 'routine housekeeping'.

Next, the huge water tank known as *The Neutral Buoyancy Trainer* which gives astronauts the sensation of what it will be like flying in space itself. A full-scale Shuttle crew cabin mid-deck, airlock and payload bay doors are completely immersed in the tank providing a simulated zero-g environment in which the trainees in their full space suits are taught the basics of extravehicular work by frog-suited instructors.

The *Shuttle Mission Simulator* – known as SMS – is perhaps the most exciting and demanding of all the training facilities. (See the later Chapter 'A Ride To The Heavens' for an account of my own experiences in one of these simulators.) For once seated inside this large, somewhat unprepossessing piece of equipment (not unlike an ungainly juke box from the outside) the trainee finds himself undergoing an unforgettable experience. His senses will tell him he is flying in space – but he will never actually leave the Houston hangar.

One of NASA's training manuals which I was given puts the purpose of this time and space machine in typically dry, scientific language. I cannot resist quoting a few lines from it:

> The SMS is a computer-controlled facility to simulate flight dynamics and the operation of all crew station displays. It is used to provide training on combined systems and flight team operations. It includes the capability to simulate payload support systems with mathematical models and remote manipulator system dynamic operations using computer-generated imagery. The SMS can also be interfaced with the Mission Control Center for conducting crew-to-ground integrated simulations.

For the uninitiated, it's an invitation to a twentieth century Magic Carpet ride!

While the SMS is obviously of greatest interest to those who actually fly the Shuttle, *The Remote Manipulator System* covers the business of carrying out actual work in space. This consists of a mock-up of an aft crew station, a payload bay and a mechanically operated arm. Using the helium-filled models of the cargo to be deployed in space (satellites, scientific experiments, etc.), the RMS can simulate payload grappling, berthing, and the operation of the Shuttle's mechanical arm.

Finally, there is the *Spacelab Simulator* (SLS) which consists of a segment of one of these 'flying' laboratories which are taken up into space in the hold of the Shuttle and there used by specialists in the crew to carry out experiments. Such missions differ from that being undertaken by the first Briton in space who will be helping to launch a satellite from the cargo hold.

It is as a result of extensive training on these pieces of complex equipment that the group of people who are to fly the Shuttle are welded into a smoothly functioning team. The team will consist of a commander, a pilot, a mission specialist and one or more payload specialists (to a maximum of seven). In present missions, the first three categories of crew members are already fully trained: being joined at a later stage by the payload specialists – that is, experts in the particular cargo being carried – who have undergone rigorous training for at least a year. One NASA wit has nicknamed the payload specialists 'The Stevedores of Space' – an epithet which seems to have stuck!

The four Shuttle-bound Britons fall into this category, and two of them will be selected for Mission 61-D and 61-H, to serve with senior NASA astronauts who have space flights to their credit. At the time of writing, the first mission in the Shuttle *Challenger* is scheduled for January 1986 and the second in *Columbia* in June of the same year.

The NASA flight manual is very specific about the 'pecking order' of the Shuttle crew during their flight (and even during the eight, half-an-hour shifts which make up their working day):

> The commander is responsible for the safety of the crew, and has authority throughout the flight to deviate from the flight plan, procedures, and assignments as necessary to preserve crew safety or vehicle integrity. The commander is also responsible for the overall execution of the flight plan in compliance with NASA policy, mission rules and Mission Control Center directives.

Above: *The exterior of the Shuttle Simulator, used to train crewmen for the Space Shuttle missions.*

Right: *A view of the open payload bay of the Shuttle* Challenger, *photographed with a camera attached to the helmet of Astronaut Bruce McCandless, space-walking by means of a nitrogen-propelled manoeuvring unit.*

The pilot is second in command of the flight. The pilot assists the commander in the conduct of all phases of Orbiter flight and is given delegated responsibilities (e.g. during two-shift orbital operations). The commander or the pilot will also be available to perform specific payload operations.

The mission specialist co-ordinates payload operations and is responsible for carrying out scientific objectives. The mission specialist will resolve conflicts between payloads and will approve flight plan changes caused by payload equipment failures. He or she may also operate experiments to which no payload specialist is assigned or may assist the payload specialist. During launch and recovery, the mission specialist monitors and controls the payload for vehicle safety.

And lastly to the humble crewman (or woman):

The payload specialist manages and operates experiments or other payloads assigned to him or her and may resolve conflicts between payloads and approve flight plan changes caused by payload equipment failures. The payload specialist will be cross-trained as necessary to assist the mission specialist or other payload specialists in experiment operation. He or she may operate those Orbiter payload support systems that are required for efficient experiment operation, as well as other systems such as hatches and the food and hygiene systems, and will be trained in normal and emergency procedures for crew safety.

The primary function of both of the British payload specialists is the launching of two new military satellites, *Skynets* 4A and 4B. Once in orbit about 160 miles above the Earth, our 'Stevedore' will oversee the launching of one of these highly delicate and technical pieces of equipment which is destined to remain in a geostationary orbit of 19,324 miles for seven years and provide instant communication between British warships, aircraft and military bases around the world.

The British-made *Skynets* will replace an earlier system, *Skynet* 11B (actually one of the first of its kind to be launched in 1974, but now no longer functioning), and provide the Ministry of Defence with an essential type of control over its far-flung forces. The necessity of this was underlined only recently during the Falklands conflict when the British authorities had to rely on American and NATO satellites for their control.

The cost of each of these launchings will be in excess of $40 million, and though the satellites manufactured by Marconi Space Systems and British Aerospace are approximately the size of a small family car, they require the most careful adjustment and checking to ensure they glide precisely into orbit and do not wander off uselessly into space. (Such a misfortune, of course, occurred to the Weststar Satellite which was so dramatically rescued by astronauts Joseph Allen and Dale Gardner from the Shuttle *Discovery* in November 1984.)

This particular mission is a doubly-challenging one for the payload specialist, for a military satellite is even more complicated than the usual communications version and contains a number of highly secret characteristics which have to be protected against any attempts to 'tap' into it – or even destroy it.

If such demands were not enough in themselves for the first Briton in space, there is also the little matter of coming to terms with *being* in space. It is perhaps no surprise, therefore, that three of the four men from whom the lucky pair will be chosen, should come from the armed forces. And the fourth man, Christopher Holmes, is a scientist and also the project manager for the *Skynet* system.

Holmes, who some believe to be the favourite for the first berth on the Shuttle, is in his mid-thirties and an expert on satellite communication. With a degree in Physics, he has a life-long fascination about space. 'When I was a teenager,' he recalls, 'I stayed up all night to watch the Moon landing.'

Commander Peter Longhurst, a few years older, and trained as a weapons engineer, has also been associated with the *Skynet* project since 1981. As Director of Naval Operational Requirements, he, too, is fascinated by the challenge of the Shuttle mission.

'Look how the American Moon landing caught the imagination,' he says,

Overleaf: *In front of the mighty* Saturn *rocket at the Houston Space Center, Texas, the first British hopefuls. From left: Peter Longhurst, Christopher Holmes, Nigel Wood, Tony Boyle. Tony Boyle was later replaced by Major Richard Farrimond. It was announced on 26 April 1985 that Nigel Wood would have the honour of being the first Briton in space.* (photo: Graham Miller)

'though it could of course all have been done without humans. I would like to see something like that on a smaller scale happening with us, to get children interested and also build up British expertise.'

This hope of raising British interest in space so as not to fall too far behind the Americans and Russians (who have put over 160 people in space between them) is also shared by the other two men, Major Richard Farrimond of the Royal Signals and Squadron Leader Nigel Wood of the RAF, both in their mid-thirties. Major Farrimond has a degree in telecommunications, while Squadron Leader Wood is a BSc in aeronautical engineering. Wood was a fighter pilot before becoming a test pilot and was actually flying some of the fastest jets in the world at Edwards Air Force Base in California when he was invited to join the other three for training. It comes as no surprise to hear the Squadron Leader admit that, 'I should really like to be *piloting* the Shuttle!'

All four men bring enthusiasm, fitness and a high degree of technical knowledge about satellites to their assignment. It is the business of learning to fly in the Shuttle and prepare themselves for work in space that has been taking up their time and energy in Houston. The size of their task is nicely laid out for them in the NASA book of rules from which I once again quote. (Also to give any would-be astronauts an idea of what would face them.)

Training a payload specialist for an Orbiter-only flight requires approximately 180 hours. A flight with *Spacelab* pallets requires 189 hours of training and one with a *Spacelab* module requires 203 hours.

Two months of nearly full-time training approximates 320 hours of available time, half of which is spent in formalised classroom and trainer/ simulator training. The remaining time at JSC can be allocated to Shuttle payload flight plan integration and reviews, flight/mission rules development and reviews, flight techniques meetings, and flight requirements implementation reviews. For some complex payloads (e.g. multidiscipline), the dedicated training may take more than 2 months.

Flight independent training for the payload specialist involves those crew tasks necessary for any crewmember to function effectively during flight; this training totals approximately 124 hours. Flight-dependent training can be divided into two types: payload discipline training, and training necessary to support Shuttle/payload integrated operations. The second is characterised by integrated simulations involving the entire flight operations support teams. Approximately 115 hours are devoted to this type of training.

And in case anyone should be in the slightest doubt about how long all this preparation adds up to, the manual concludes: 'The payload specialist training may start as long as two years before the flight.'

It would, though, be quite wrong to think that NASA has closed the door to space flight to all but the most intelligent and highly trained people. Indeed, at Houston visitors are encouraged if they so desire to 'sign up' for future Shuttle missions! And it makes no difference what your age, sex, weight or even nationality might be. For the US government has instituted a 'Citizens in Space' programme which entitles anyone to put down their name for a journey to the stars in any one of the existing Shuttles – *Columbia*, *Challenger* and *Discovery* – or even the soon-to-be-ready *Atlantis*, due to make its maiden voyage in September 1985. Though I'm not the first, there is now another addition to the list of prospective 'Writers in Space' . . .

Although it is true that the first Briton will be viewing the Earth and the rest of the galaxy from space a long time after a good many other people from other nations, at least the moment has come and at last the dreams of a number of earlier Britons who all made important contributions to the ultimate conquest of space have been realised. I mean people like the Wiltshire monk who was the first man to fly, and the Yorkshire landowner who built the first successful heavier-than-air flying machine. Or more recently the Science Fiction writer who invented the satellite and the research team who first thought of a reuseable spacecraft. These, and others, had dreams which, had they all been followed through, *might* have enabled a Britain to be the *first* man in space rather than following behind, courtesy of American technology.

But as we stand on the threshold of this achievement there is no time for recriminations: neither, though, should we let the remarkable achievements of our pioneers be forgotten. And so in the pages which follow, due recognition is accorded not only to the great American and Russian space achievements, but to the very real contributions of Britain as well.

An oblique view of a portion of Great Britain looking northeastwards across England and Wales, taken from Skylab 3.

42

A DREAM OF STARWALKING

DESPITE THE undeniable fact that the first Briton in space is following well behind the American and Russian astronauts, the British have an important place in the history of the conquest of space: a fact that is often overlooked and rarely given full credit in any discussion. The evidence for such a claim is not hard to discover, either, and several groups of modern investigators have gone to considerable lengths to substantiate the truth about claims of flight made in days long since past. The almost fabled achievements of pioneer 'airmen' have, in fact, been recreated in modern times and proved beyond all reasonable doubt to have *succeeded*.

It is now established, for instance, that the first man to have been *proved* to have flown in the air was an English monk named Eilmer about the year A.D. 1010. It is equally true that the inventor of the first heavier-than-air machine to leave the ground was a Yorkshireman named Sir George Cayley in 1853. And it was an eighteenth century English rocket expert, Sir William Congreve, who added fins to his war rockets and took the first step towards the modern spacecraft; just as it was a British respiratory physiologist, Dr J. S. Haldane, who outlined a design for the world's first space suit early this century. And we must not forget that it was English minds who devised the first communication satellites, proposed a lunar spaceship fifty years ago which was almost identical to the *Apollo* spacecraft which carried out the actual landing, and even thought up the Space Shuttle years before the Americans. But for the indifference of the authorities these pioneers might well have enabled Britain to lead in the space race.

Of course any discussion about a British contribution to space flight must, regardless of chronology, begin with a mention of one man, Sir Isaac Newton (1642–1727) who has been called 'the greatest scientific genius of all time'. For it was this Lincolnshire-born scientist and mathematician who by defining the laws of gravity and inertia of the solar system established the basic principle upon which space flight is founded. He also gave mankind a key to understanding the Universe by first dividing sunlight into its spectrum of colours, thus conceiving the spectroscope, one of the most valuable tools of astronomy. And if this was not enough, he also explained the lunar motions.

Small wonder, then, that his spirit should pervade the eventual conquest of space. In his memory, the units of thrust needed to launch a space vehicle have been called 'newtons'. For example, 9.8 newtons will just balance one kilogram of weight, while 10 newtons will lift it. To raise the Space Shuttle off the launch pad requires about 30 million newtons.

On a more human and certainly most appropriate note, his name was featured during the first manned rendezvous with the Moon. Just 241 years after his death, in December 1968, as *Apollo 8* left its Earth orbit and was coasting in a precisely determined trajectory towards the Moon, one of the astronauts, Bill Anders, had this conversation with his five-year-old son observing the flight from Mission Control in Houston.

'Who's driving, Daddy?' the boy asked. 'Is it you?'

'No,' came the reply.

'Is it your friend, Colonel Borman?'

Again a negative. 'Then it must be Captain Lovell?' The child persisted.

Once more Bill Anders had to say no. Then who *was* driving, the little boy asked, almost exasperated.

The astronaut was silent for a moment and then his voice came distinctly over the emptiness of space.

'Sir Isaac Newton,' he said. And all those who heard him knew he was right.

Mankind's dream of starwalking, of escaping from his natural environment and soaring like the birds into the heavens can, though, be traced back to antiquity, and indeed the aspirations of many unknown 'birdmen' are embodied in the Greek legend of Daedalus and his son Icarus and their flight from captivity on the island of Crete. Ovid's account of the pair's ill-fated journey with wings made from feathers, linen fastenings and wax is beyond question imaginary, although scholars do agree that it expresses in story form 'one of man's oldest desires' and certainly stimulated many of the subsequent attempts at flight.

It was almost certainly not until about A.D. 1010 that a man flew in anything like the manner of Daedalus and Icarus. This 'birdman' was a remarkable fellow named Eilmer, a Benedictine monk who lived in the Wiltshire Abbey of Malmesbury. The naturally fortified hill-top town of Malmesbury had become a special religious teaching centre under the Saxons, drawing to it monks whose principal occupations were copying documents about agriculture and crafts. Among their number was Eilmer, a man said to be particularly interested in scientific matters. He lived between the years 980 and 1080, and the basis of the events that made him famous are to be found recorded in the classic work, *De Gestis Regum Anglorum* written by William of Malmesbury. William (c. 1095–1143) is widely regarded as the best informed and most reliable historian in twelfth century England, and as he lived in the self-same Abbey as Eilmer and at no great interval after the events he describes, his words give added authenticity to the monk's achievement. This is his description of the events that changed history:

He (Eilmer) was a man learned for those times, of ripe old age, and in his early youth had hazarded a deed of remarkable boldness. He had by some means, I scarcely know what, fastened wings to his hands and feet so that, mistaking fable for truth, he might fly like Daedalus, and collecting the breeze on the summit of a tower, he flew for more than the distance of a furlong. But, agitated by the violence of the wind and the swirling of air, as well as by awareness of his rashness, he fell, broke his legs, and was lame ever after. He himself used to say that the cause of his failure was his forgetting to put a tail on the back part.

In the years since, students of flight have attempted to establish just what kind of equipment Eilmer might have devised for his flight, and the general concensus of opinion is that they were most likely to have been rigid wings of considerable size attached to his back and arms. The distance he flew, according to William, was 'spatio stadii et plus' which translates as 'more than

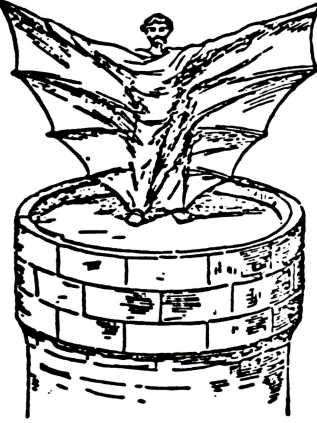

the distance of a stadium'. As a Greek stadium was 600 feet in length we therefore have a reasonably accurate measurement.

A recent and most enthusiastic investigator, Max Woosnam, thinks it is unlikely that Eilmer flew from the spire of the Abbey as legend says, but rather from the side of the steep hill on which Malmesbury Abbey stands. Apart from his study of the original records, Mr Woosnam has also reached a conclusion about the size of the monk's wingform and tested his theory by re-enacting the flight in 1962.

He told me how he pieced the clues together. 'I do not think the Abbey in Eilmer's time was any higher than the remnants of the present one, which was built 150 years later,' he said. 'This height is about 100 feet and was about 120 feet to the base of the original spire. However, the spire fell down in the fifteenth century and the base was damaged. The spire was an extra 300 feet making 400 odd feet to the top, but I don't believe Eilmer climbed that.

'So in 1962 I made a re-enactment with an RAF Royal Tournament Display Team who flew off the present Abbey on a cable. I made the costume and wings and they were only 24 square feet. Even so, it was awkward standing on top of the Abbey in a breeze. I have more recently designed and flown lightweight Delta kites of 250 square feet with a base of 25 feet. If Eilmer took off from 120 feet high and flew 660 feet on a 5.5 glide angle, I

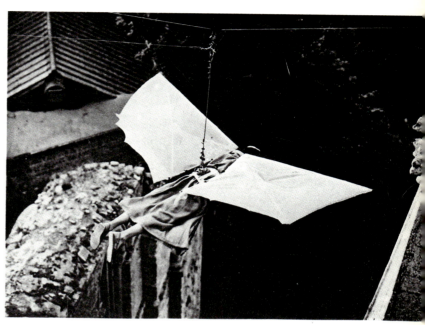

Above: *The mechanically assisted re-enactment of the flying monk's spectacular first flight shows him* top left *on the roof* top right *'gliding' between the spires, and* above *landing in the churchyard.*

Opposite: *The flying monk, Eilmer, 'collecting the breeze on the summit of a tower'.*

believe that he must have used at least 200 square feet of wing area considering the state of aerodynamical knowledge at that time.'

I find Mr Woosnam's evidence for the success of Eilmer's flight most convincing – and it is worth recording that the RAF team staged a second re-enactment in 1976, and there are plans for further trials in the near future. On what we know now, I believe Eilmer deserves wider fame than the simple stained glass window that records his endeavours in Malmesbury Abbey.

A greater credit for his part in the conquest of the air is also due to the Yorkshire squire, Sir George Cayley, who proposed and flew the first practicable powered aeroplane in the year 1809. According to the aviation historian Charles Dollfus, Cayley was 'the true inventor of the aeroplane – the most powerful genius in the history of aviation'. Not only did Cayley improve on the system of lift into the air as sought by Eilmer and the other pioneers, but he proposed a means of *thrust* as well.

Cayley, in fact, launched the world's first successful man-carrying aeroplane across the valley of Brompton Dale, near Scarborough, in 1853, a full half a century before the Wright Brothers took off at Kill Devil Hills in America. Though history has given scant regard to this achievement, Wilbur Wright was generous in *his* praise when he spoke about Cayley in 1909.

'About one hundred years ago,' he said, 'an Englishman, Sir George Cayley carried the science of flying to a point which it had never reached before and which it scarcely reached again during the last century.'

The main reason for the public ignorance of Cayley's triumph was the isolation in which he worked, and the fact that his notebooks and sketches were only rediscovered by chance in 1927, some seventy years after his death. Cayley was actually born in 1773, a decade before the first hot-air and hydrogen balloons were invented in France. Raised on an education of science and mechanics, he became absorbed in the idea of creating heavier-than-air machines when still only a teenager. Then, at 18 years of age, he inherited his father's baronetcy, and the wherewithall to persue his ambition on the family estate of Brompton Hall.

In 1796, Cayley's notebooks show, he invented what was to all intents and purposes a model helicopter, using whalebone as a spring and feathers as rotor blades. His life-long study of the flight of birds also lead him to draw up plans for a powered monoplane, and at the turn of the century he wrote:

The stained glass window in Malmesbury Abbey showing, on the right, Eilmer, holding his wings.

I am well convinced that Aerial Navigation will form a most prominent feature in the progress of civilisation during the succeeding century. . . . I feel perfectly confident that this noble art will be brought home to man's general convenience and that we shall be able to transport ourselves and our families, and their goods and chattels, more securely by air than by water, and with a velocity of from 20 to 100 miles per hour.

By creating a model of his design, Cayley had invented nothing less than the modern aeroplane in embryo – but he knew that there was suspicion, even fear, among people about his experiments, and so he shrouded his work in secrecy. In the years which followed, he pioneered the concept of the powered fixed wing aircraft, the semi-rigid airship, the principle of streamlining, and the idea for an aero engine which he called a 'First Mover' that would 'weigh less than a man.' Then in 1804 he flew an actual model of a glider-like aeroplane, complete with kite-type wing and adjustable tail unit.

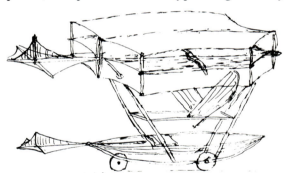

Cayley's original design of the fixed-wing glider with wheeled undercarriage which was the first heavier-than-air flying machine.

'It was very pretty to see it sail down a steep hill,' he observed, 'the least inclination of the tail towards the right or left made it shape its course like a ship by the rudder.'

Four years later Caylcy had come up with a hot-air engine which, although it proved impractical for flight, was enthusiastically embraced by industry. His further observation of bird flight had also made him realise the importance of cambered wings: a factor which has subsequently proved so vital in the design of jet aircraft.

A DREAM OF STARWALKING

He worked on relentlessly and in 1808 drew up the first plans for a tension-sprung wheeled undercarriage to enable his flying machines to take off and land. A year later – still only 35 – Cayley built and flew his first ever full-size glider.

The busy Squire saw this as the first major step towards his ultimate objective of an aeroplane that would carry people. Making more tests with the glider, he confided in his notebook: 'When any person ran forward in it, with his full speed, taking advantage of a gentle breeze in front, it would bear upwards so strongly as scarcely to allow him to touch the ground; and would frequently lift him up and convey him several yards together.'

Inexplicably, on the verge of his greatest triumph, Cayley gave his attention over to civil matters, and for almost the whole of the next thirty years did no practical experiments and only produced the occasional article on aeronautics. However, one of these did lay down the principle of jet propulsion as a power source, while another on aerodynamics has since been said to contain 'the foundations upon which the whole vast science of flying is founded.' It was, in essence, a resume of all his work at Brompton Dale.

It was not, in fact, until he was 70 that Cayley again devoted himself whole-heartedly to aviation matters, and in 1853 he built what was to prove to be the first man-carrying aeroplane to achieve free flight.

It was an odd-looking craft, to be sure. A huge kite was attached to what appeared to be the hull of a yellow dinghy mounted on bicycle-type wheels. Yet, curious as it may have seemed, on a warm July day in 1853, the machine was launched into the air from Brompton Dale with Cayley's terrified young coachman as its only passenger. For 400 yards the craft sped across the valley, to land with a bump on the far side. A delighted Sir George found the trembling coachman still clutching the sides of the machine. 'Please, Sir George,' the fellow whispered, 'I wish to give notice. I was hired to drive and not to fly!'

Those words – as significant in their way as those of Yuri Gagarin in space or Neil Armstrong stepping onto the Moon – were the first to be uttered by the first man to have actually flown. Tragically, neither Sir George Cayley or history has left us a clue to the name of that pioneer airman.

The Cayley flying machine never took to the air again, and Sir George died in December 1857, leaving no obvious trace of his achievements save a single glider that served for a time as a hen roost on the estate! His diaries, as I mentioned, lay hidden in an attic for seventy years.

It was not until some years after the discovery and publication of the Cayley notebooks and diary, that Anglia Television decided in 1972 to reconstruct the flying machine and see if it *would* actually fly. They called in Commander John Sproule of the Fleet Air Arm to supervise the construction and lavished £2,500 on the 39 foot craft with a 30 foot wing span and 460 square feet of sail made of finely-woven poplin. The finished machine, exact to every one of Cayley's specifications, weighed just 200lbs.

After tests at Lasham Gliding School in Hampshire where it was built by a team led by Ken Fripp, Anglia transported the flier to Brompton Down and recruited Lasham's chief instructor, Derek Piggott, a former RAF pilot, to take the place of the nervous coachman.

Piggott approached the flight with his usual equanimity. He had, after all, flown some of the finest stunt sequences in films such as *The Blue Max* and *Those Magnificent Men in Their Flying Machines*, and having studied the design plans felt the machine *ought* to fly. But it was the location at Brompton Down that bothered him.

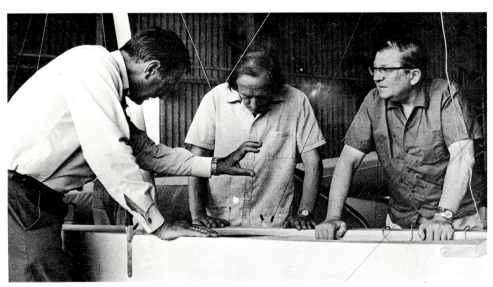

A DREAM OF STARWALKING

Left: *At work on the replica of Cayley's flying machine. Left to right: Ken Fripp, builder; Derek Piggott, pilot; and Commander John Sproule who realised Cayley's drawings.*

Top: *Derek Piggott takes the place of Cayley's coachman as the replica is prepared for take-off.*

Above: *The first test flight of the replica at Lasham Gliding School in Hampshire where it was built.*

'At first I thought it wouldn't be safe to do it,' he recalls, 'because when you go off the top of one hill it looks certain you are going to fly into the side of the other. But after a day or two I decided it was worth trying, though I really did think I was going to get hurt. I can understand how the coachman felt – he must have been horrified!'

And so with the television cameras ready to record the success – or failure – of Piggott's flight, he was towed across the ground and into the air. 'It was tremendously exciting,' he says, 'The machine proved airworthy and reasonably stable, flying in a graceful hop. I managed to land, after a 15 second flight of about 200 yards, at the bottom of the valley.

'This was really the first aircraft a man actually flew in, so to be the second man in history to fly the world's first aeroplane – that's fun!' Piggott added.

At the end of the experiments, Ken Fripp commented happily, 'My feeling is that we have flown the replica enough to be able with confidence to give Cayley, who was way ahead of his contemporaries, the credit due to him.' Commander Sproule saw even further possibilities in the craft. 'I have no doubt,' he announced afterwards, 'that a small modern aero engine – the light and powerful 'First Mover' that Cayley yearned for – would get our replica off the ground into sustained flight.'

A DREAM OF STARWALKING

Left: *After construction at Lasham, the glider was transported to the site of the original flight, Brompton Dale in Yorkshire. Test pilot Derek Piggott is seen here operating the rudder.*

Top: *The replica of Cayley's flying machine coming in to land.*

Above: *A rather bumpy landing at the bottom of the valley at Brompton Dale -- but a successful second flight in the world's first aeroplane.*

Though such a proposal has not yet been forthcoming, the experiment has had the effect of making Cayley's achievement better known. A full length feature film on his life and work is being planned, and as a result of some vigorous lobbying, Sir George has been posthumously elected to the International Aerospace Hall of Fame in San Diego, California, thereby being acknowledged for his contribution to aeronautics and, in turn, space flight. Beside a model of his machine are quoted some of his words which seem to echo across the years as a challenge which is at last being answered by the conquest of space:

'An uninterrupted navigable ocean, that comes to the threshold of every man's door, ought not to be neglected as a source of human gratification and advantage.'

In looking through Cayley's papers, one might almost be tempted to suggest that he also outlined the idea for a spaceship as well as aeroplanes! For among his designs is one for a *gunpowder-powered* craft! The Yorkshire baronet was clearly aware that a superior form of power would be needed to drive one of his machines above anything other than a nominal height, and consequently picked on gunpowder as the most powerful 'driver' he could think of. Sadly, he took his idea no further than the sketchy drawing (illustrated on these pages), which he captioned 'Design for a Gunpowder-Engined Monoplane Model' and dated 1850.

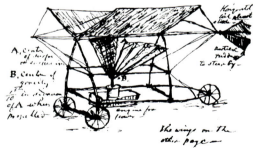

Cayley's design for a gunpowder-engined monoplane model which he drew in 1850.

It is quite likely that Cayley had been attracted to the idea of using gunpowder by the work of another powerful genius who had been busy in England at much the same time, Sir William Congreve (1772–1828). Just as Cayley's work marked the true beginnings of practical aviation, so Congreve proposed the most significant development in rocketry which led ultimately to the space rocket. It was this London born scientist who introduced the idea of 'fins' on rockets – the first step towards a controlled and directed flight.

Although gunpowder had been used as a weapon in warfare for centuries, it was the threat of an invasion of England by Napoleon Bonaparte in the early years of the nineteenth century that inspired Congreve, the inventive son of a Royal Artillery officer, to develop a rocket weapon that could be targetted on the enemy fleet across the Channel. Previously the firing of such 'war rockets' had been haphazard, intended mainly to frighten the enemy.

Working at the Royal Laboratory in Woolwich, Congreve came up with the idea of a long stick mounted on the side of the rocket to stabilise it in flight. These rockets were duly put into operation in two Royal Navy missions to Boulogne in 1805 and 1806 and accurately wreaked havoc on the French forces. Congreve, however, was not entirely happy with the effectiveness of the 'guiding stick' (it's effective range was between 900 and 2,700 metres depending on size) and had the brilliant idea of a series of fins around the base of the rocket. At a stroke he gave the rocket absolute stability and controllable targetting – only one step removed from the axially rotated rocket which has conquered space.

Although the system of axel rotation was first developed in America in 1815, Congreve produced a number of rockets with similar characteristics at this time, their rotation being imparted by an offset thrust, and in them can be seen several of the elements adopted into the sophisticated machines which 150 years later carried men into space. Why, even the appearance of one of Congreve's Incendiary Rockets – which could carry a warhead of 50 lbs of gunpowder for up to 2,500 yards and which can be seen in the National Air and Space Museum in Washington alongside modern *Viking* and *Saturn* space rockets – bears a striking resemblance to modern rockets in its appearance and the shape of its nose cone.

On his death in 1828, Congreve in fact left a number of more advanced designs for rockets up to eight inches in diameter and weighing as much as 1,000 lbs. He had also begun work on a rocket which contained a series of charges, each ignited successively, thereby providing a 'booster' to its flight duration. Here, again, he had correctly foreseen what was to prove another important element in space flight.

Although it is undeniable that the first person to seriously suggest that the rocket could be used as a means for interplanetary travel was the great Russian, Konstantin Tsiolkovsky (see Chapter 5, 'The Sputnik Legacy') in 1903, my research for this book has brought to light a startlingly accurate prophesy about this in an obscure little book published in Britain some *four years earlier*. This virtually-unknown footnote to space history, is made all the more tantalising by the fact that we have no idea whatsoever who the author of the book was!

The work is called *Half Hours in Air and Sky*, and it was issued in 1899 by the small London publishing house of James Nisbet & Co. As the title indicates, it is a popular survey of the heavens based on the then-known facts about the planets of the solar system and the universe in general. Although, as I say, the author is anonymous, he gives every indication of being a scientifically-minded individual for his accurate definition of rocket theory is enough to have made even his distinguished contemporary, Jules Verne, green with envy! (To his credit, of course, Verne did anticipate many of the elements of space flight in his book, *From The Earth To The Moon* (1865) right from blast-off to splash-down complete with the problems of weightlessness.)

'In the infancy of physical science,' our unknown prophet writes, 'it was hoped that some discovery should be made that would enable us to pay a visit to our neighbour, the Moon. The only machine independent of the atmosphere we can conceive of, would be one on the principle of the rocket. The rocket rises in the air, not from the resistance offered by the atmosphere on its fiery stream, but from internal reaction. The velocity would, indeed, be greater in a vacuum than in the atmosphere, and could we dispense with the comfort of breathing air, we might with such a machine transcend the boundaries of our globe and visit the other orbs.'

How remarkably accurate those words still seem a century on – all the more so when you consider how even today the principle of rocket motion is *still* often misinterpreted!*

* It is also interesting to note in the light of these remarks, a comment by the veteran British rocket enthusiast, A. V. Cleaver that by 1900, 'some 90 per cent of the *science* required to go to the Moon was known, but the engineering and technology base required had not yet been acquired.'

A DREAM OF STARWALKING

If the problem of how to survive in space without air seemed insoluble to the unknown author, it was, however, on the verge of being solved by a quietly dedicated English scientist named J. S. Haldane. For the beginning of the development of the pressure suit – which ultimately became the space suit – was very much absorbing the time of this respiratory physiologist, and in 1907 he developed a method of decompression by stages which made it possible for deep-sea divers to ascend to the surface safely and avoid the ever-present risk of getting the 'bends'.

Four years later in 1911, Dr Haldane had extended his experiments to another of the elements – air – and proposed the use of an oxygen pressurised suit for ascent to high altitudes. By 1933, he had perfected his theory to such a degree that he had designed and made the first pressure suit ensemble: a modified version of the diving suit.

That same year, a 27-year-old American experimenter named Mark Ridge asked Haldane for a pressure suit that he could use in the open basket of a high altitude balloon he was preparing to fly. In November, he came to London to test the suit in a special altitude chamber and, sealed inside, duly reached a simulated altitude of 90,000 feet. There he remained with no ill-effect for 30 minutes. (It is worth mentioning that without the suit at that altitude, the water vapour in Ridge's blood vessels would have begun evaporating rapidly because of the low atmospheric pressure and he would have died.)

This experiment proved to be the first successful test of the world's first space suit – the essential item of space travel which also made possible a landing on the Moon and will surely facilitate similar journeys to other worlds. And, once again, the invention was that of an Englishman.

It was perhaps appropriate that the British Interplanetary Society should have come into existence in the October of this same year, and although the Society is neither the oldest nor largest such group in the world it is highly regarded: particularly in America at the National Air and Space Museum where a notice declares for all to see that, 'Today the British Interplanetary Society remains one of the most important learned societies in the world devoted to space flight.'

High praise indeed for an organisation that began in the most modest circumstances in Liverpool – yet several of whose members were later to make significant contributions to space history.

Although there was undoubtedly a certain amount of interest in space travel in Britain during the early part of this century, the major handicap for anyone who wanted to conduct actual rocket experiments was the stringent Explosives Act of 1875, sometimes sarcastically called the 'Guy Fawkes Law'. This forbade all private experimentation and manufacture of 'gunpowder, nitro-glycerine, blasting powders, coloured fires and rockets'. Yet some more determined souls were not to be thwarted in their ambitions, and, for instance, during the early 1920s a Yorkshireman named Ernest Welsh of North Ferriby made a number of test flights of a rocket he claimed could climb to a height of five miles. Although Mr Welsh was evidently inspired with the same spirit of adventure as his fellow Yorkshireman, Sir George Cayley, he saw his rocket very much as a military weapon which could be used against enemy aircraft by throwing out a shower of molten metal pellets as it climbed into the air. Although the British, French, American and even Russian authorities were said to have shown interest in this 'Death Rocket', nothing further was heard of it after some test flights made for the British Army in the summer of 1924.

While Ernest Welsh clearly had no thoughts about space travel, a reclusive gentleman named Harry Grindell-Matthews conducted a number of secret rocket experiments behind the barbed-wire fences of his large estate at Mynydd-y-Gwair in Wales. He apparently had his sights more definitely fixed on a projectile which could reach the stratosphere, but was also hoping to produce an 'aerial torpedo' for use against hostile aircraft. A combination of a lack of skilled labour, a failure to interest the authorities in his ideas, and the eventual exhaustion of his finances, brought an end to Grindell-Matthews's tests in the late Thirties.

The only successful use of a rocket in Britain at this time appears to have been carried out under the auspices of the Post Office who allowed a young German rocket expert, Gerhard Zucker, to run trials for a 'Rocket Mail' on the Sussex Downs in the summer of 1934. Several rockets containing mail were fired and successfully reached their destination at Brighton on the coast. The Post Office were, seemingly, initially unaware of the Explosives Act, and once this was brought to their attention, Britain's first rocket mail service came to a speedy end.

A DREAM OF STARWALKING

It was against this background of generally less-than-successful experimentation, however, that a young structural engineer from Wallasey near Liverpool named Philip Cleator formed The British Interplanetary Society, 'for the stimulation of public interest in the possibility of interplanetary travel, the dissemination of knowledge concerning the problems which the epoch-making achievement of an extraterrestrial voyage involves, and the conducting of practical research in connection with such problems.' Cleator had been fascinated by rockets since his childhood and this interest had been heightened by reading the American 'Science Fiction' magazines then beginning to cross the Atlantic. Shortly after his twentieth birthday, he inherited his father's engineering business which gave him the laboratory and equipment to further his experiments with rockets. Then, in 1933, after reading a statement by W. A. Conrad of the United States Navy, that it might soon be feasible to reach the Moon by rocket, he decided to form the Society.

Cleator's plan came to the ears of a *Daily Express* reporter, whose paper considered the idea newsworthy enough to be splashed on the front page of the September 8 issue. The result was a small deluge of mail which convinced the young engineer that there were a good many other people like himself in the country. Among the early members were the Science Fiction writer, Olaf Stapledon; Professor A. M. Low, the pioneer of radio controlled guided missiles in World War I and a long-time advocate of space flight (who later became the Society's President), and a 19-year-old Somersetshire farmer's son whose eventual fame was to overshadow all the rest, Arthur C. Clarke.

Because of the legal restrictions on experimentation, much of the BIS's work took the form of theoretical articles published in a monthly *Journal* which ranged across a variety of topics from Rocket Cars to highly innovative plans for Repulsor Rockets. All bore the hallmark of dedicated scientific research and clear thinking. Philip Cleator also wrote the very first British space book, *Rockets Through Space*, which was published in 1936 and did much to stimulate public interest in the subject. (The release that same year of an excellent film version of H. G. Wells' classic novel, *Things To Come*, went still further in generating interest.)

The single most important project of the Society was, though, a space ship to travel to the Moon. Conceived and designed by several leading members of the group between the years 1937 and early 1939, it was eventually described and blue-printed in the BIS *Journal* just before the outbreak of World War II. It remains a tantalising thought – what might have become of this perfectly feasible plan had not that terrible war intervened? What does remain a fact is that the American aerospace company, Grumman, borrowed the idea when making the actual Lunar Module which eventually put a man on the Moon. Though the BIS lander differed in shape to the Grumman design, the principle was the same. A two-stage craft would land on the lunar surface, and when it was time to depart, the top, ascent stage would lift off using the descent stage as a launching pad. It was certainly a radical departure from all previously conceived ideas of what a spaceship should look like.

As the BIS's own diagram shows, the spaceship was divided into six main steps, the higher ones being smaller than the lower ones, each divided longitudinally into a total of almost 2,500 asbestos-bound cellules. This cellular construction enabled the propellant tanks to be jettisoned as soon as they became empty. The larger steps at the bottom of the rocket also contained fewer cellules than those higher up. Each cellule was a solid propellant rocket motor complete with fuel and fired electrically. And when fired, each burned to the exhaustion of its propellant with its thrust determined by its initial design. The ending of the thrust allowed the cellule to become unhooked and drop off. Because of the large number of cellules, control of the total thrust could be obtained by selecting the time and rate of ignition of the cellules. Solid propellants, it was believed, also gave a high density and good structural strength to the ship.

The overall dimensions of the BIS spacecraft were to be 105 feet tall by 19 feet 7 inches in diameter. Its total weight would be 900,000 kg, consisting of 1,150 tons of propellant; the pressure cabin for two or three crew members weighing in at three quarters of a ton; five tons of batteries for ignition and other power needs, food stores for 20 days (the estimated time of the mission), tools, water and air; and the general structure another 275 tons.

Because of the stipulated weight limit of one ton to enable the lunar lander to make its journey from the Earth to the Moon and back, the actual design proved extremely taxing. In the end, the design team came up with the shape of a hemisphere on top of a downward pointing cone. The addition of an upward pointing cone below this formed the barrier between the life

JOURNAL

OF THE
British Interplanetary Society

JANUARY 1939

6d. to non-members.

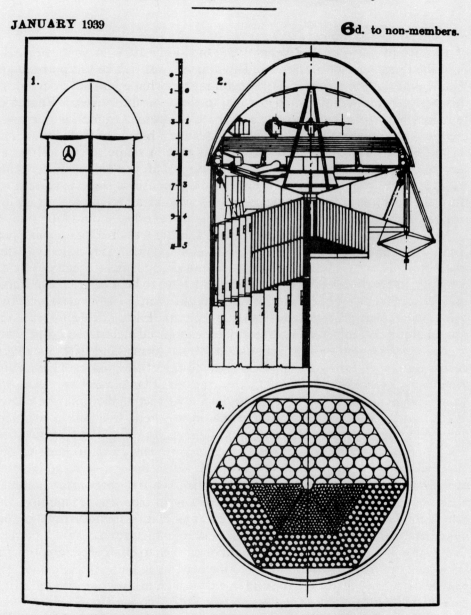

Title page of the BIS Journal for January 1939, showing the design for a lunar spaceship.

compartment and the first layer of propulsion cellules. And in the space between these two cones were to be located air locks, air conditioning plant, heavy stores, batteries and the liquid fuel for manoeuvring the craft. To protect the crew from the high temperatures that they might expect to experience, the life compartment had a paraboloid reinforced ceramic carapace mounted over the hemispherical nose.

The tricky problem of landing on the Moon's surface was solved by providing the lunar lander with a multi-legged system linked to hydraulic shock absorbers. The actual descent would be controlled by systematic firing of the rockets – the self-same rockets which would later facilitate the take-off and return to Earth.

This remarkable prototype not only established all the basic principles which are still accepted today, but also pointed out precisely the developments in technology which were required to make a Moon voyage possible. Furthermore, the proposal went into considerable detail about what would be required by the crew members on such a mission. It specified the kind of high protein foods needed, the electrically heated ovens and aluminium or plastic utensils to facilitate eating. It outlined the rubberised, heat-resistant space suits that would have to be worn and the kind of on-board exercises the crew might perform to keep fit. It listed balsa wood pencils, light-weight playing cards, razor blades and various items for sanitary needs. And for use on the Moon were specified: flat-bottomed shoes, signal rockets, dark goggles and sunburn lotions, dynamite charges for removing large lunar rocks, specimen tubes and reagents, a powerful telescope, a microscope, a spring balance and gravity pendulum, a cine camera and an ordinary camera, and so on. No item was too insignificant to be listed, no detail too small to be considered. The BIS proposal even added that crew members should be given a supply of various kinds of money 'with which the intrepid explorers will pay for their return to civilisation should they land in one of the more barbarous regions of Earth'.

Despite the undoubted potential of their plan, Philip Cleator and the members of the BIS were unable to convince the authorities to take the matter seriously – even at all. With war clouds gathering over Europe, they possibly did not help their cause by pleading that a mission to the Moon could probably be undertaken for the same cost as building one naval destroyer!

A DREAM OF STARWALKING

With a resigned air – and in words that were to prove uncannily prophetic – Cleator wrote to a friend in America in September 1939, 'Others are trying to interest the Powers That Be, but for myself I have given up – though it would certainly be ironical if a world war were required to demonstrate the potentialities of the rocket. Somehow I hate to think of the device I have fathered for so long turned to destructive purposes.' (It was, as we shall see, rocket expertise, developed by the Germans during the late 1930s and up to 1945 which ultimately helped put both the Americans and the Russians into space.)

When, 30 years later, in July 1969, the *Apollo 11* mission to the Moon took place just as the BIS had outlined, one of the most interested observers was a leading member of that ground-breaking team: Arthur C. Clarke. And he was able to recall with a rueful smile those pre-war years and how in an article about the British project entitled, 'We Can Rocket To The Moon – *Now!*' published in the Summer 1939 issue of *Tales of Wonder*, he had said, 'The flight to the Moon, so far from belonging to the realms of fantasy, has been brought down into the realm of contracts and estimates; and, indeed, it should not cost much more than £250,000, due to the fact that most of the ship's components can be mass produced. The first lunar expedition, therefore, will leave the Earth as soon as sufficient backing can be found for it: the plans – and the men – are ready, *now!*'

Sadly, of course, the fulfilment of this dream lay three decades and millions of dollars away . . .

The spaceship was not, in fact, the only important piece of space technology in which Arthur C. Clarke was involved – indeed he can be credited with the invention of the whole idea of communication satellites. While it is true to say that today he is the best known SF writer in the world, and the film based upon his story, *2001: A Space Odyssey* (1968) provided a whole generation with its image of space exploration, such accolades tend to overshadow his perhaps equally important earlier *practical* work.

Born in Minehead in 1917, he was fascinated by space from his childhood and at school was nick-named 'Spaceship'. His early career was in the Civil Service working in the Exchequer and Audit Department, but all his spare time was taken up with the BIS and writing. During the war he served with distinction in the RAF – rising to the rank of Flight Lieutenant – and in the

It was thirty years after the BIS design was published before the dreams of those pioneer designers became reality with the first moon landing.

immediate aftermath took a first class honours degree in physics and mathematics at King's College, London. When, in 1946, the BIS was reconstituted, Clarke was elected Chairman, and quickly proved himself one of the leading thinkers on interplanetary matters. In the pages of the Society's *Journal*, for instance, he published a whole series of articles on the problems that would have to be overcome to achieve interplanetary flight. And in 'The Dynamics of Space Flight', for example, he precisely defined the speeds a rocket would need to achieve to escape the Earth's gravitational pull and the effects upon a spacecraft and its occupants once they were beyond the influence of the mother planet.

But perhaps more important than all these was a brief article entitled 'Extra-Terrestrial Relays' which he published in the magazine *Wireless*

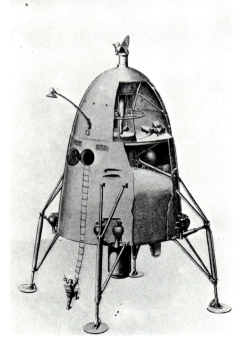

Design of a lunar spaceship from Arthur C. Clarke's 1951 publication, The Exploration of Space.

A DREAM OF STARWALKING

World in October 1945. For, in the space of a few thousand words he created one of the technological marvels of the century: the communication satellite.

It is remarkable to reflect that this piece appeared *twelve years* before the Russian sputnik was launched in October 1957 and a whole *seventeen years* before *Telstar*, the first television satellite which went into space in July 1962! I think it is fair to say that if Clarke had patented this idea instead of selling it to the magazine for a fee of just £15 he might be an even richer man than he already is. He can, though, comfort himself by the thought that his definition of the conditions ensuring a satellite remains stationary over a given point on the Earth have been named after him – 'Clarke's Orbit'.

During the course of this extraordinary article – which he prefaced with the remark that 'many may consider the idea proposed too far fetched to be taken seriously' – Clarke discussed how a long-range rocket developed from the German war-time V2 type could break free from the Earth's gravitational pull and would then become an artificial satellite circling the world, 'like a second Moon, in fact'. He went on:

> There are an infinite number of possible stable orbits, circular and elliptical, in which a satellite would remain if the initial conditions were correct. One orbit, with a radius of 42,000 km has a period of exactly 24 hours. A body in such an orbit, if its plane coincided with that of the

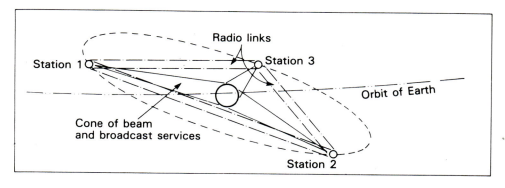

From Clarke's article in Wireless World, *this design shows how three satellite stations would ensure complete coverage of the globe.*

Earth's equator, would revolve with the Earth and would thus be stationary over the same spot on the planet. It would remain fixed in the sky of a whole hemisphere and unlike all other heavenly bodies would neither rise nor set.

It could be provided with receiving and transmitting equipment and could act as a repeater to relay transmissions between any two points on the hemisphere. Moreover, a transmission received from any point on the hemisphere could be broadcast to the whole of the visible face of the globe. A single station could only provide coverage to half the globe, and for a world service three would be required, though more could be readily utilised (see diagram). The stations would be arranged approximately equidistant around the Earth.

It is very largely due to this article and the subsequent development of the communications satellite which it prompted, that the term 'Global Village' is now no longer a figure of speech. For the satellites circling our planet do provide instant communication from one side of the globe to the other, as Clarke visualised. And it is satisfying to be able to add that although Britain may lag behind in some areas of space technology, we are the leading developers of satellite and communications facilities, and apart from the Americans are the foremost user of satellites. (For the record, the first joint UK and USA satellite, *Ariel 1* was launched in 1962, the same year as *Telstar 1*, the first TV communications satellite linking Europe and America. The first all-British satellite, *Ariel 3*, was put into orbit in 1966).

Tucked away in this same article, Arthur C. Clarke also prophesised another step in the conquest of space. 'Using material ferried up by rockets,' he wrote, 'it would be possible to construct a "Space Station" in such an orbit. The station could be provided with living quarters, laboratories and everything needed for the comfort of its crew, who would be relieved and provisioned by a regular rocket service. This project might be undertaken for purely scientific reasons as it would contribute enormously to our knowledge of astronomy, physics and meteorology.

'Although such an undertaking may seem fantastic,' he adds, 'it requires for its fulfilment rockets only twice as fast as those already in the design stage.

Opposite: *Launching of the UK-X-4 Satellite in March 1974.*

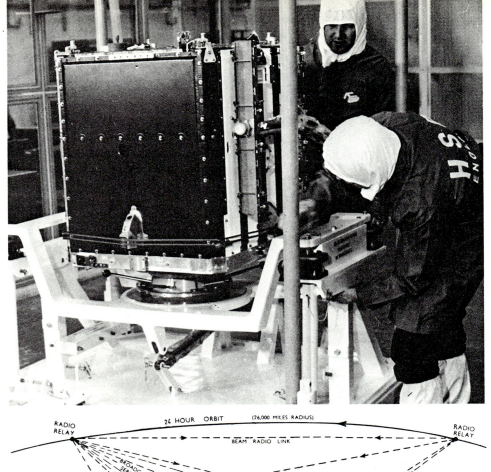

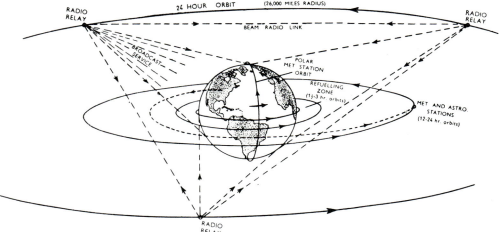

Top: *The spacecraft for the United Kingdom's X-4 mission is shown being prepared for systems test and launch checkout prior to its launch on 8 March 1974.*

Above: *From* The Exploration of Space, *published by Clarke in 1951, this is his design of the orbits of a space station.*

Opposite: *Arthur C. Clarke's most remarkable idea for the future – a Space Elevator, a giant cable link from Earth to a satellite in geostationary orbit 22,320 miles above.*

Since the gravitational stresses involved in the structure are negligible, only the very lightest materials would be necessary and the station could be as large as required.'

Here again Clarke has provided the outline for what is now one of the American's major space projects. But his thoughts on the future do not end there, and while I was working on this book he came up with his next remarkable idea for the future – a Space Elevator, a giant cable link from a point on the Earth's equator to a satellite 22,320 miles above.

'It's perfectly feasible,' he says. 'We already have much of the technology required. With it people and equipment could be hoisted up by purely mechanical means such as speed capsules moving inside the cable, thereby reaching orbit without any use of rocket power. From the space platform they could be transported to other destinations or into orbit. I imagine that by the end of this century people will be thinking seriously about building a Space Elevator – and then constructing it in the next.'

The chances are, of course, that this brilliant idea will be taken up by someone other than ourselves and prove yet another missed opportunity just like the most recent and perhaps saddest of all in the catalogue of unfulfilled British contributions to space travel. For just as Arthur C. Clarke created the idea of satellites to be developed by the Russians and the Americans, so a

team of British scientists who actually designed the first Space Shuttle have had to watch others perfect their ideas.

The design for the first reusable space orbiter was completed in 1964 after several years of work by a team at the British Aerospace's research unit at Warton Aerodrome on the edge of the River Ribble in Lancashire. Heading this group was an aeronautical engineer with the very English name of Tom Smith. He and his group – which varied in size from eight to 30 people – came up with the plans for a shuttle that might well have flown twenty years before the now-famous group of NASA Shuttles, *Enterprise* (the prototype), *Columbia*, *Challenger*, *Discovery* and *Atlantis*.

Some experts believe that the plans for a shuttle which Tom Smith and the others drew up called 'Project MUSTARD' (Multi-Unit Space Transport and Recovery Device) are still superior in some ways to the present American version, but this is not something Tom himself would claim. Instead, he talks in a very matter-of-fact way about the giant leap for British space travel which never quite came off.

'It all started in the early sixties when the Ministry of Aviation gave the British Aircraft Corporation a contract to study Hypersonics,' he recalls. 'To do this, the Corporation formed a team at Warton and we began by looking at things of the Concorde type. From there we went to high speed aircraft which would travel at Mach 12 – and ramjets as well.'

As a result of this study, the team came to the conclusion that it would be possible for an air-breathing craft to go into space – but first they had to determine what the payload was likely to be and particularly how much it would all cost. 'You have to remember that rockets then had chemical propulsion and were one-off shots,' says Tom. 'It cost a great deal to get something into orbit, and most of the components were simply thrown away in doing so.'

With costs established as the major factor, Tom knew that reusability was of prime importance. 'Using the airbreathing designs we had and adapting them for space was like trying to turn a submarine into an aeroplane, so we knew that was out,' he continues, adding that as the team knew it was vitally important to get out of the atmosphere very quickly they looked at launching a second stage rocket from a modified aircraft. But as this would be both costly and inadequate from a payload point of view, a vertical take-off lifting

74

body seemed to be the answer.

'The first design we came up with was for something with two stages, one stacked on top of the other,' Tom says. 'The lower stage was the booster and the second went into orbit. But two different designs for the two stages made it expensive and the lower stage was difficult to control for recovery.'

Undeterred by this seemingly impossible problem, the team then threw all the traditional ideas to one side and came up with the proposition of putting three, arrow-shaped units side-by-side. They would all take-off vertically together: two of the units being rocket boosters and the third the crew-carrying section which would go into orbit.

The launch and flight pattern of Project MUSTARD, the British Shuttle designed by Tom Smith, resembles that of the US Shuttle which it preceded.

'The advantage of this,' Tom explains, 'was that you could have all three motors firing at once for a powerful and efficient take-off, and with three identical components costs would be lower. Fuel could also be carried in the two boosters and pumped through to the orbiter so that on separation it could still carry a full fuel load even though its rockets had been operating from take-off.'

According to the plans for the British shuttle, the orbiter unit would separate from the two booster units after a flight of about 150 seconds when it had reached an altitude of 30 miles. Refiring its own rockets, the spacecraft would achieve its predestined orbit at 1,000 miles just over eight minutes later. Re-entry to the Earth's atmosphere would begin at 5,000 miles from the landing point, with the peak heating and maximum G Force of 5.1 occurring at 270 miles from base. At touchdown the craft would be travelling at just 100 knots.

After the initial separation, it was planned that the two booster sections would be guided back to base by remote control, so that along with the piloted orbiter they could be re-used in future missions. The orbiter's tricky re-entry into the Earth's atmosphere with its attendant danger of over-heating and burn-up was to be achieved with the same kind of angled manoeuvring that later proved so successful with the NASA Shuttles.

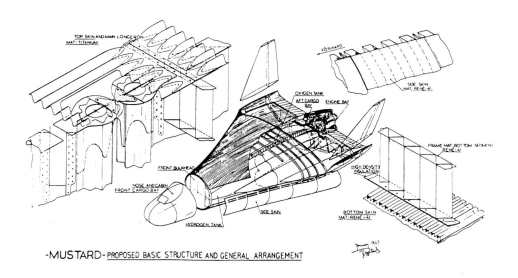

-MUSTARD- PROPOSED BASIC STRUCTURE AND GENERAL ARRANGEMENT

Sadly, this pioneer British Shuttle was destined to go no further than the design stage.

A DREAM OF STARWALKING

By the time the specifications had been completed and the spacecraft visualised by an artist (see examples here), Tom Smith and his team were convinced that the MUSTARD shuttle would break even on its costs at between 50 to 100 missions. But sadly this pioneer British vehicle was destined to go no further than the drawing board.

'I think part of the trouble with the shuttle being dropped was that it was so far ahead of its time,' Tom reflects today. 'There was a lack of political will in the UK in particular, as well as Europe in general. Of course several years later the Americans started working towards their own shuttle design and . . . the rest is history. There's nothing worse than being right at the wrong time!'*

Reflecting on information such as this, one cannot escape the conclusion that if Project MUSTARD had been allowed to proceed with government backing and funds, a Briton might well have been the very first man in space a decade or more ago rather than following in the footsteps of others. And what a fitting climax that would have been to the brave achievements of Eilmer, the monk of Malmesbury; the redoubtable Sir George Cayley; Sir William Congreve; Arthur C. Clarke and all the others who have dreamed of escaping from the confines of Earth when such thoughts were just dreams.

In the light of what the future promises in space, is it not time for the British government to reconsider its attitude – as the *Sunday Express* urged only recently?

'For the first time in centuries,' the newspaper said, 'the British have failed to place themselves in the forefront of voyages of exploration. Real space enthusiasts reproach British governments for dithering and for failing to create industrial infrastructure for the future. The rocket programme fizzled out years ago with the scrapping of Blue Streak. And although there is limited co-operation with European projects, no sign has emerged of plans for a more full-bloodied programme.'

---

* Ironically, while no trace of Project MUSTARD remains in Britain, a model of the orbiter has a place of honour in the National Air and Space Museum in Washington. The small, white model of the orbiter complete with its two booster units which once stood in Tom Smith's Warton Office, now occupies a glass case alongside an exhibit explaining the development and success of the NASA Shuttles. Though the importance of the British invention may be lost on the average visitor, it is a salutory lesson to anyone from these islands.

Above: *An artist's conception of the manned, wingless space vehicle which was the first US design for a reusable shuttle.*

Top Right: *The next stage in US Shuttle development was this wingless M-2 research vehicle, pictured here in 1963.*

Bottom right: *A 75-ton simulator of the first US Shuttle which was used for weight and lift-off experiments at the Kennedy Space Center, Florida.*

It *is* time for Britain to have its own Space Agency so that the man who travels on the NASA Shuttle is not the first of a few but the first of many. Surely our tradition demands it!*

* Since I completed this section, it has been announced that the British Government has at last responded to pressure and plans to set up a National Space Centre – a kind of 'mini-NASA', probably based at Farnborough in Hampshire – to co-ordinate research. In making this announcement, Information Technology Minister, Geoffrey Pattie, admitted that Britain's political commitment to space had not been on a level with that of British industry. He added, 'The responsibility for space has been scattered and there is clearly a need for a much sharper focus for Britain's space effort.'

The Space Shuttle Enterprise, *mated to an external fuel tank and two solid rocket boosters prior to take-off.*

A RIDE TO
THE HEAVENS

THE SPACE SHUTTLE which carries astronauts to and from the wide open spaces which lie beyond the boundaries of our world, is undeniably one of the wonders of modern scientific technology – a reusable orbital transportation system able to fly as many as 100 missions and combining in one machine three remarkable capabilities. It can be launched like a rocket, manoeuvred like a spacecraft and landed like an aeroplane. Ten years in the researching and development at a cost of $10 billion, it is the first winged vehicle to have gone into orbit and also to have achieved orbital re-entry. And though it weighs 100 tonnes and can generate a thrust of 470,000 lbs from each of its three main liquid oxygen fueled Rocketdyne engines (called SSMEs) and fly at an orbital speed of 17,600 mph, the sensation of travelling in it has been described by one astronaut as 'like being in a gondola under a hot air balloon racing across the sky'.

This marvel of human ingenuity is just 122 feet long and is launched into space on the back of two Thiokol solid rocket boosters (SRBs) attached to a Martin Marietta external fuel tank (ET) containing hydrogen and oxygen. The two boosters shut down and separate from the orbiter at a height of 30 miles, while the external tank parts company after the craft's main engines cut off as it reaches maximum speed at an altitude of 75 miles. Three to four minutes later, at 115 miles high, the two orbital manoeuvring rockets (OMS) in the tail – fueled by Monomethyl Hydrazine which can generate a thrust of 6,000 lbs each – boost the spacecraft into its designated orbit. Any small velocity changes and altitude adjustments which are later required can be achieved by the group of 44 small engines situated around the Shuttle's nose cone (called RCSs) which are also fueled by Monomethyl Hydrazine. (The two boosters parachute down to the sea and can be recovered and reused, although the external tank burns up during re-entry into the Earth's atmosphere.)

The Shuttle has an altitude range of between 115 and 600 miles, and a special thermal protection system of over 31,000 individual silica fibre tiles measuring six inches by six inches. These cover almost three quarters of the surface area and protect it and the crew from the extremes of high and low temperatures which they must face during the mission – particularly during re-entry. Because of the rather brick-like appearance of these tiles, some NASA insiders have been heard to refer to the Shuttle rather tongue-in-cheek as a 'Flying Brickyard'!

The crew of any Shuttle can number up to seven: a Commander, pilot, a mission specialist, and up to four payload specialists – the 'Stevedores of Space' I referred to earlier. The craft is capable of carrying payloads weighing up to 65,000 lbs, and this can include communication and tracking satellites, space telescopes, probes, orbiters and landers – even a Teleoperator Retrieval System, more commonly known as a 'Space Tug' used in the launching and retrieving of space hardware. Perhaps most important of all is the Spacelab: a complete laboratory for the astronauts to work in along with a series of pallets on which measuring instruments can be mounted and experiments exposed to space. The Spacelab is primarily concerned with studying the Earth, the atmosphere, the sun and planets, as well as conducting biological tests and ultimately making preparations for the building of a manned Space Station. (The success of processing biological and chemical substances in the pure and weightless conditions of space looks likely to have far-reaching implications to the manufacturing industry on Earth, particularly that of pharmaceuticals.)

To facilitate their work in space, the astronauts are provided with Extravehicular Mobility Units (space suits to you and me) which enable them to function beyond the confines of the Shuttle, and Manned Manoeuvring Units complete with thrusters, power outlets, lights and cameras, enabling them to move freely in space itself.

But what is it actually *like* flying in the Space Shuttle? Being part of a

Top: *This astronaut's eye view is taken from the pilot's seat in* Columbia *prior to launch.*

Above: *Two astronauts sit in the cockpit of the Shuttle Simulator during training at the Johnson Space Center.*

million-mile space odyssey that has at last turned man's long-held dream of travelling to the heavens and back into a reality? By talking to some of the astronauts who have done so, and taking a 'ride' in the $60 million Shuttle mission simulator at the Johnson Space Center in Houston, it is possible for even an earth-bound writer to quite accurately recreate this almost magical journey.

What one has to appreciate right from the start is that the Shuttle is not 'flown' in the accepted sense of the word. The pilot does not *fly* this complex spacecraft in the way any normal aircraft – be it a light plane or a mighty jet airliner – is flown. Much of its operation is controlled by electronic systems, and though it would be impossible to launch and land the Shuttle without the aid of its giant computer system, it *is* provided with the traditional stick and rudder pedals so that the astronaut-pilot could hand-fly it when it is in orbit or in the final stages of landing. It is also true that for the last half an hour of its return to base, the Shuttle becomes a glider (albeit doing Mach 25 as it re-enters the atmosphere at 400,000 feet) and finally touches down at about 180 knots – a flight which some aeronautical experts believe a skilled pilot could undertake manually as long as he paid the most careful attention to the command bars. Even taking your eyes off these bars for a moment to glance out of the panoramic spread of eleven cockpit windows as the world rushes by at a speed of 26,000 feet per second is enough to cause disaster – in the simulator at least, as I know to my cost! (This crucial half-hour in the flight plan is, incidentally, described in NASA terminology by the term EI, for 'Entry Interface' and WOW, meaning 'Weight On Wheels'!)

Before the Shuttle begins the final count-down to take-off, there are no less than five days of pre-checking that all systems are operational. However, the crucial last seven minutes are these:

Seven Minutes. The access arm to the Shuttle is withdrawn leaving the crew finally isolated and lying at 90 degrees on their backs in the spacecraft. (They actually enter the Shuttle an hour and fifty minutes before blast-off.)

Six Minutes. The pilot runs through a multi-item checklist on his three special cathode ray tube displays, prior to starting the engines.

Five Minutes. The Commander and pilot activate the orbiter's flight instrumentation and are free to monitor the system status and prepare for lift-off.

A RIDE TO THE HEAVENS

Four Minutes. The Shuttle's aero control surfaces are automatically run through a moving profile to ensure they will work correctly on the ascent and are also ready to fly a re-entry should the flight be aborted for any reason.

Three Minutes. The orbiter is switched to its own electrical cryogenic fuel cell system, and the main engines are *gimbaled* to demonstrate their readiness.

Two Minutes. The external tank liquid oxygen and hydrogen vent valves are closed to build up pressure in the tank.

One Minute. The Shuttle's five interlinked general purpose computers take over control of the countdown sequence. The ground launch sequencer remains on line, however, supporting and monitoring the launch.

Thirty Seconds. The sound suppression system floods the flame deflector trench beneath the pad with thousands of gallons of water to prevent accoustic or heat damage to the rear of the Shuttle as it rises.

Ten Seconds. The five computers send a start command to the three main engines, the SSMEs. The first engine ignites at 6.45 seconds, followed by the other two at 120 millisecond intervals. All reach 90 per cent of thrust.

Three Seconds. The crew feel a distinct lurch forward and backward as the SSMEs cause the orbiter to be thrust in the direction of the external tank. This movement of almost 19 inches is known as the 'Twang' and it has to be allowed to cease in order that the solid booster ignition will strike when the craft is once again in the upright position.

Zero. The SSMEs having fired for approximately six seconds, the computer ignites a 20 foot sheet of flame down the centre core of each solid booster which provides explosive lift-off energy and thrusts the Shuttle upwards from the launch pad. The cumulative energy generated by the five rocket motors at this moment is equal to 7.5 million pounds.

As the solids ignite, the Shuttle Commander calls, 'Lift-off!', and the mission to the stars has begun.

Just how the real thing affects a pilot has been described by one of NASA's most experienced men, Captain Robert L. Crippen, who flew the *Columbia*, the very first shuttle to go into space, in April 1981. This is how he vividly recalls those moments.

'When the Shuttle's main engines went off we heard a kind of bang in the cockpit,' he said, 'setting off some vibration – but nothing very significant.

We could still read all the instruments and tell that the vehicle was rocking. We had been expecting to get a really swift kick in the pants, but it seemed the solid fuel or the structure itself absorbed some of the shock.

'At lift-off, I had been expecting something like an aircraft carrier hydraulic catapult shot,' the former Navy pilot continued, 'where the g-force loads really spike, almost knocking the breath out of you. But what we felt was more like a steam catapult shot where we had a pretty heavy vehicle and a nice steady push. Still, there wasn't any doubt we were taking off!'

The next nine minutes of the Shuttle's flight as it climbs from zero speed to an astonishing Mach 25, shedding its two booster rockets and fuel tank in the interim, are superlative examples of precise space technology. Though the orbiter's computers automatically handle the operation, there are only fleeting moments for the crew to watch space looming up in front of them: they must keep a careful check on the control panel for any deviation in the craft's functioning. (As a matter of interest, there are 'cue cards' pinned up around the cockpit with information to assist the Commander and pilot in any emergency during this critical period of ascent.)

But let me recreate these minutes in a little more detail. As the Shuttle actually lifts off the launch pad, its first movements are lateral as well as vertical, and it in fact shifts a few feet as it aligns its thrust vector under its centre of gravity. By the time it has cleared the top of the tower, in just under eight seconds, the alignment has been corrected and the slight vibrations have ceased.

Almost immediately, at an altitude of 400 feet, the Shuttle commences a 120 degree roll and pitch manoeuvre which is designed to put it in an inverted position for the ascent. Climbing rapidly to 8,000 feet in 32 seconds, the three main engines are throttled down to 65 per cent of their thrust to allow the Shuttle to go through Mach 1 speed. This is achieved at 24,000 feet, just 52 seconds into the mission, and then the engines are returned to maximum power. A minute later and the Mach 3 barrier registers on the airspeed indicators as the craft roars up to 120,000 feet.

Opposite: *This spectacular shot shows the Space Shuttle* Columbia *moments after its powerful solid rocket boosters were ignited.*

Settling into the flight pattern now, the crew next await the separation of the two solid rocket boosters, which are thrust free at a height of 27 nautical miles where the atmosphere is thin but still existent. This parting just over two minutes into the Shuttle's journey is accompanied by a blast noise and a bright flash seen through the cockpit windshield.

Two more minutes on, and the flight reaches perhaps its most critical moment. Now flying at 63 miles high at Mach 6.4, the Shuttle has achieved enough power to begin the process of going into orbit – or if there are any signs of malfunction to turn around and return to base. Once committed to the looping arc which ultimately puts the craft into orbit, however, there is no going back – for one revolution of the Earth at least.

Attaining an altitude of 70 nautical miles, the Shuttle goes through its biggest manoeuvre, dropping from an angle of flight 19 degrees *above* the horizon to 4 degrees *below* it, the velocity screaming up by one Mach every 30 seconds.

This spectacular dive gives the upside down astronauts a breath-taking view of the Earth, but there is little time to appreciate it, for having attained a maximum speed of Mach 25 after just eight and a half minutes of flying, the Shuttle is now ready for the engines to be shut off and the external tank jettisoned. Computer failure here could lead to disaster for the craft is under its heaviest g-pressure and the astronauts would find any kind of manual control extremely difficult. The Shuttle parts from its last remaining accessory just as nine minutes is registered on the display consol. This parting is accompanied by the sound of a mild boom which has been described rather aptly by one astronaut as sounding like 'a cherry bomb firework exploding'.

The next four minutes are busy ones as the Commander and the pilot oversee the functioning of bursts of power by the manoeuvring rockets to get the Shuttle into its correct orbit. The OMS rockets perform this task in virtual silence and there is little more sensation in the craft than that of a gentle push. The climb which eventually levels off into the orbit takes about 30 minutes.

It it important that the pilot remembers before settling into orbit to close the doors through which the pipes linked to the external tank run. Should this sensitive area of the craft be left open by mistake, re-entry into the Earth's atmosphere would be extremely dangerous, if not actually impossible.

Once again, just as we are fortunate to have one astronaut's description of the moments of the Shuttle's lift-off, so another has provided an eyewitness account of what it is like to be in space and in orbit. The man is Joseph Allen who flew in *Columbia* on its fifth mission in November 1982, making 82 orbits and covering two million miles.

'The first day in orbit you're like a baby deer on ice,' Allen, a payload specialist says. 'Your feet go out from under you. You bang into everything, and you move around with your arms held out in front of you. Space is like an undersea world, with a dream-like quality in which things seem to move more slowly.'

Then, becoming more specific about this sensation which, of course, the astronauts are carefully prepared for in the huge simulation tank back on Earth at the Johnson Space Center, Allen continues, 'With weightlessness, there's no such thing as right-side-up and upside-down. There's a ceiling. There's a floor. There are walls. Sometimes it's convenient to have your feet on the ceiling with your head towards the floor, but you never think the floor is the ceiling. You're upside-down, but you don't feel any different.

'To begin with, you push off to go to the other side of the spacecraft, and you wind up not going where you want to go. Once you're on your way, though, you can't stop until you get to the other side. You could be headed for a wall where things are attached by Velcro. Of course, you put a hand out to stop yourself, but you do it awkwardly and bump all the things off the wall. You adapt to being weightless by trying different ways to move around, realising little by little that all it takes to get from one place to another is a *gentle* push.'

Allen said he found eating and drinking in space a most peculiar experience. Though food tasted the same as on Earth, nothing dripped from it. And with liquids everything had to be drunk through a straw with a clip on it to keep the liquid from continuing to run out after each sip!

Like all astronauts, Allen was fascinated by the absolute peace and quiet of space – the total silence that makes it almost impossible to comprehend that the Shuttle is travelling at 17,500 miles an hour. Even when he went outside the craft to help put two satellites into orbit, he found no sensation of movement and not even a whisper of the sounds the Shuttle and the satellites make in the normal atmosphere of Earth.

There is, though, a spectacular sensation of movement given by the speed

Above: *A view of the Earth's horizon from the Shuttle* Challenger. *Through the clouds can be seen the English Channel.*

Opposite: *A back-up payload specialist, Robert B. Thirsk, during zero-G training aboard NASA's KC-135 aircraft. This aircraft is used for training and exposing Shuttle crewmembers to weightlessness.*

with which sunrise and sunsets occur as the Shuttle hurtles around the Earth. Sixteen examples of each occur during the course of a 'Shuttle day', and they have eight different bands of colour ranging from scorching red to the deepest blue. Night also falls with such amazing speed and intense blackness that astronauts can only tell where the Earth is in relation to their craft by studying the stars until they, too, stop – which is the point at which our planet blocks out the light.

'I had a hard time getting over the way you see the Earth,' Allen recalls. 'It's no longer flat as you see it from an ordinary aeroplane – it's a globe. You know the Earth is round because you can see the roundness. And then you realise there's another dimension because you can see layers as you look down. You see clouds towering up and their shadows on sunlit planes. You

see a ship's wake in the Indian Ocean and bush fires in Africa – even the reds and pinks of the Australian desert. I was just awestruck by it all.'

But after the magic of riding in the heavens, there eventually comes the little matter of returning to Earth. And as one NASA official put it to me, bringing the Shuttle into the atmosphere at around 400,000 feet while still doing the Mach 25 which it has maintained during its orbits, is 'rather like riding a bronco for the first time – one slip and you're off.' And coupled with the actual speed of the craft, is the return of gravity and the effect *that* will have on the astronauts after the time they have spent in zero-g. (They have, in fact, to put on pressurised flight suits to cope with this change.) Perhaps not without justification, a returning Shuttle has been described as 'a cannonball with wings.'

To facilitate the re-entry descent, the Shuttle has to make a number of hypersonic S-turns to dissipate its speed while at the same time preserving the energy, and also reshape its orbit so that the lower part comes down and just touches the uppermost reaches of the atmosphere. The effects of this manoeuvre can be quite spectacular as Commander John W. Young, the man in charge of the first *Columbia* Shuttle mission, and a veteran of several Apollo missions including a landing on the Moon, has described.

'I was expecting to see some bright re-entry colours,' the man widely regarded as the world's most experienced astronaut recalls. 'At a velocity of about 24,500 feet per second we started to get a very light pink glow around the side windows, which gradually built up and covered all the forward windows and tended towards orange a little bit up near the nose. I don't see how you can have a pink glow outside your spacecraft and not have 2,000 degrees Fahrenheit. So if you look at the nose and it is orange then it has got to be 3,000 degrees Fahrenheit! It sure was pretty, though!'

Another dramatic change also occurs on the control panel on the g-force dial where the needle has for so long been stationary at zero. Suddenly, it moves forward and the astronauts are conscious of sounds once again. There is a roar that builds up with increasing fury as the air of the atmosphere reverberates around the Shuttle.

It is essential now for the Commander and pilot to ensure that the craft does not pitch up and become a kind of giant air brake, for such a drag would cause it to almost instantly burn up despite the heat tiles. So a controlled sideslip is required, and at 300,000 feet and doing Mach 24, the craft is rolled onto an eighty degree bank, holding its own angle at between 28 and 40 degrees. There are further roll manoeuvres performed at 260,000 feet, 215,00 feet, 115,00 feet, and 85,000 feet by which time the speed has been steadily and safely dropped to just below Mach 3.

'From about Mach 3 on down there was never any doubt we were slowing,' Commander Young recalls of his first mission, 'because from a psychological standpoint we were also slowing down like crazy. Then somewhere a little bit above Mach 2 we started to get a little transonic buffet, a little washboard kind of effect. When the shock waves move over a spacecraft it's bound to bump you a little bit since this is not exactly a streamlined aeroplane.'

It is not, in fact, until the Shuttle is finally dropping below 3,000 feet at a speed of about 280 knots that the speed brakes are applied. By then, the flight plan is nearing completion, and a further slowing to about 180 knots gives the ideal touch-down speed. The craft lands in the traditional aircraft style, rear wheels first and then gently tipping forward onto its single nose wheel. Although this landing is computer controlled like all the other elements of the orbiter's flight, all the pilot's skill and training needs to be alert for there is still the chance of a last-minute disaster until the machine runs to a halt on the home-base tarmac.

And so another journey has come to an end. But whether you are stepping out of an actual Shuttle – as those fortunate astronauts have been doing with increasing frequency in recent years – or are merely exiting from the authentic cocoon of the flight simulator in Houston, this is an experience to change a person's life.

For the NASA Space Shuttle has finally made a space age vision come true.

A RIDE TO THE HEAVENS

Opposite: *Astronaut Dale A. Gardner, left, holds a For Sale sign, making light reference to the status of the recaptured communication satellite, Westar VI, during* Discovery's *retrieval mission in November 1984.*

Below: *Completing the first full test of the Space Transportation System, the Shuttle* Columbia *is seen here on its final approach prior to landing at NASA's Dryden Flight Research Center, Edwards, California.*

Bottom: *Space Shuttle* Challenger *landing on the 15,000 foot Shuttle Landing Facility at the Kennedy Space Center, Florida, on 19 November 1984.*

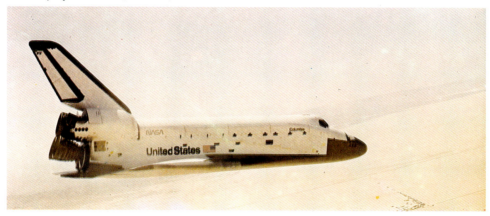

The Army-JPL Bumper Wac, *made up of a V-2 and a* WAC/Corporal, *lifts off from White Sands Proving Ground, New Mexico.*

THE WINGS
OF THE EAGLE

JUST OVER 25 years ago, in October 1958, the National Aeronautics and Space Administration organisation was born under commission from President Dwight D. Eisenhower. Created out of the old National Advisory Committee for Aeronautics (NACA), America's now-famous space agency was destined to be responsible for some of mankind's greatest achievements in space – not the least of these putting the first man on the Moon.

But the birth was not an easy one. The three services, the Army, Navy and Air Force, who had been jointly involved in NACA were far from keen that the space programme to which the organisation was dedicated should be handled by a civilian organisation. And though there had been some hiccups in rocket developments which had allowed the Russians to steal a march and launch *Sputnik* in 1957, the 43-year-old NACA had a record of solid achievement behind it. What it didn't have was the broad charter the new era demanded.

It had, in fact, been the First World War in Europe which had brought about the birth of NACA, when the American authorities had become aware of just how primitive and unorganised their aeronautical establishment was. This resulted in the setting up of the Committee in 1915 to 'direct and conduct research and experiments in aeronautics'. Theoretical research was co-ordinated across the nation, and a practical laboratory was built at Hampton, Virginia, named after the American aeronautical pioneer, Samuel P. Langley. Here the world's first wind tunnel was created and in the years which followed produced important developments in streamlining and increasing the speed of aircraft.

However, it was the Second World War which really established the importance of NACA. Exhaustive research by the organisation provided manufacturers with the necessary details to build some of the finest fighter aircraft of the war – including the *Corsair, Wildcat* and *Hellcat*. And as a direct result of the impressive new technology spawned by the conflict – albeit much of it was for destruction – it was obvious that there were immense possibilities to be pursued in such fields as atomic energy, jet engines and giant rockets. All pointed to a future way beyond the confines of current aeronautics. In 1945, the atomic age, the jet age and – not far beyond – the space age, all beckoned.

It would, though, be wrong to think that the inspiration for the conquest of space came solely through the lead of NACA – America in fact had a lone experimentor named Robert Hutchings Goddard (1882–1945) who had been working towards just such a goal for over half a century.

Goddard, a reserved but immensely dedicated scientist, was born in Worcester, Massachusetts. As a teenager he developed a passion for space which consumed his life, and at as young an age as 17 he was already speculating with considerable accuracy on the Earth's upper atmosphere and interplanetary space. An entry in his diary for October 19, 1899, for instance, includes a comment about the feasibility of 'using rockets as a means of carrying measuring instruments into the atmosphere and space beyond'.

In 1907, a year before he graduated from Worcester Polytechnic Institute, Goddard wrote a remarkable essay in which he suggested that the heat from radioactive materials could be used to expel substances at high velocities through a rocket motor, thereby providing sufficient power to make interplanetary travel possible. Amazingly, this prophetic article which might well have advanced the birth of the space age by many years – not to mention America's part in it – was refused publication by all the leading scientific journals. Goddard was, quite simply, forty years ahead of his time.

He was not a man easily deterred, however, and by 1911 had received his Ph.D. and joined the faculty at Clark University. Here he taught students and worked on his passion for space travel. Soon, he was declaring that high energy propellants such as liquid oxygen and liquid hydrogen would generate the velocity necessary for interplanetary flight, and speculating that any

Earth escape rocket attempting to make such a journey would have to be constructed on the multiple, or step, principle so that the final rocket carrying the payload had sufficient terminal velocity.

During initial tests with a small rocket in 1914 using a smokeless powder along with oxygen and hydrogen, Goddard achieved exhaust velocities as high as 7,964 feet per second. He also proved for the first time through experiments carried out in a steel chamber that a rocket motor would work with greater efficiency in a vacuum than in an atmosphere. And in a report he prepared later on his work, he also estimated that a rocket weighing as little as eight to ten tons would be capable of escaping from the Earth's gravitational pull. These were significant discoveries that were to have far-reaching effects.

In the course of America's involvement in World War I, Goddard was rather side-tracked into experiments on the military possibilities of rockets, but again came up with a revolutionary design for some short and long range missiles, including a type of infantry rocket projectile which was clearly a forerunner of the Second World War 'Bazooka'.

In 1919, Goddard published a paper entitled, 'A Method of Reaching Extreme Altitudes' which did much to rekindle interest in rocket research. It also came to the attention of a German rocket pioneer, Herman Oberth, who was so impressed that he used it to develop his own theories in a book, *Die Rakete zu den Planeternraumen* (The Rocket Into Interplanetary Space) published in 1923. This in turn led to the formation of a German group, the *Verein für Raumschiffahrt* (Society for Space Travel) which attracted to its ranks several other important rocket pioneers including Wernher von Braun and Willy Ley. These men were not only to create the deadly V-2 rockets of World War II (the Vengeance Weapons, to give them their full name), but also play a vital role in the later American space achievements. Once again, we have an example of the range of Goddard's influence.

But Goddard was a prophet without honour in his own land and some newspapers who were sceptical of his work dubbed him 'The Moon Professor'. His cause was not helped by an extraordinary report which appeared in some journals on August 5, 1924 that he had actually sent a rocket to the Moon! Many people were taken in by this totally erroneous report, and poor Goddard even found himself receiving applications from

people who wanted to make the voyage. Small wonder, then, that he tended thereafter to be even more secretive about his experiments.

In reality, Goddard had only constructed and briefly experimented with a small rocket propelled by liquid oxygen and petrol. This he followed in 1926 with a second rocket using an oxygen pressure-fed system. Launched on March 16 at Auburn in Massachusetts, the ten foot high rocket made of thin sheet steel was airborn for two and a half seconds, covering a distance of 184 feet at a speed of 64 mph. It amounted to the world's first flight by a liquid propellant rocket.

The next step in Goddard's plans was to fulfil that boyhood ambition of sending an instrument-carrying rocket into the sky: and this he achieved three years later in 1929. Unhappily, the rocket plunged back to earth causing consternation among the people of Massachusetts who believed it to be a crashing aircraft, and Goddard had the utmost difficulty in allaying public fears. Thereafter he had to carry out all his launches from military firing ranges – primarily the Mescalero Ranch at Roswell in New Mexico.

Achievement followed achievement in the ensuing years. In April 1932, Goddard fired the world's first gyro-stabilised missile which landed intact thanks to the use of a parachute fitted to slow the descent. Then came bigger and more powerful rockets – the best achieving speeds of over 700 mph and flying to heights of over 7,500 feet. By the time the Second World War was

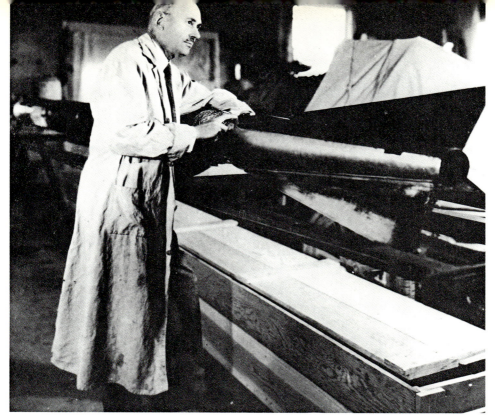

Above: *Dr Robert H. Goddard, the 'Father of American Rocketry', with his rocket in his workshop at Roswell, New Mexico, October 1935.*

Opposite: *Goddard (second from right) with one of his later designs, a four-chamber rocket, at the launching site at Roswell, 7 November 1936.*

under way he was on the verge of still greater things – but tragically ill-health began to interrupt his work and on August 10, 1945 he died after a throat operation.

But Goddard's influence did not die with him. The inspiration he had given to space-minded members of the public had resulted in the formation of the American Rocket Society in 1930, though he himself refused all invitations to become involved. This brain-child of a group of New Yorkers, led by David Lasser who was to write the first English language work on the subject, *The Conquest of Space*, in 1931, did rather more promotion work for the idea of space travel than actual experimentation, although there were a few rocket flights under its auspices, including one in November 1932 (an ill-fated attempt at Stockton, NJ) and another in May 1933 at Staten Island when a small projectile rose to approximately 250 feet in just two seconds. From these modest beginnings, the American Rocket Society was later merged with the Institute of the Aerospace Sciences which became the American Institute of Aeronautics and Astronautics in 1963, now regarded as one of the foremost technical societies in the world.

Goddard's influence was also evident at NACA, which in the immediate post-war years was picking up the challenge of the recently developed technology. In 1945 work recommenced on sending recording instruments ever higher into the stratosphere, and one of the slim, three-finned projectiles developed for military use called a *WAC-Corporal* soared over forty miles high.

In 1945, too, another rocket experimenter joined the NACA team, Wernher von Braun, the former member of the German Society for Space Travel, who was subsequently to be at the very heart of the American leap into space. Von Braun (1912–1977), a passionate devotee of rocketry since his childhood in Germany, had been suborned into the Nazi development of the V-1 and V-2 rockets at Peenemunde. Always hopeful that his work on these giant rockets would ultimately lead to the conquest of space – 'It seemed to me that the funds and the facilities of the Army were the only practical approach to space travel,' he wrote years later – von Braun surrendered to the US Army at the end of the war and travelled to America. Here he took out US citizenship and entered the employ of NACA, working for them at both the Langley Research Centre and the first rocket firing base established at White Sands in New Mexico. Here he refurbished and flew V-2's and from them developed the mighty rockets which became the core of the US space programme.

Wernher von Braun's profound knowledge of rocketry and his vaunting ideas made him a leading figure in this programme, and ultimately he became director of the George C. Marshall Space Flight Centre in Alabama. It was his *Redstone* rocket that put the first two Americans in space, and his mighty *Saturn* rocket which launched the *Apollo* programme and fulfilled his own personal dream of putting a man on the Moon. It is a fact, too, that he *could* have put an American satellite into orbit before the Russians, had he been given the go-ahead, and he was also one of the first scientists to outline a practical design for a space station. To many people, von Braun is the father of the US space programme, and certainly there can be no denying that his V-2 was the 'embryonic space ship'.

Opposite: *The* Mercury-Redstone 2 *rocket, designed by Wernher von Braun, is shown in its service stand at Cape Canaveral, Florida. This 1961 flight carried a chimpanzee to test the spacecraft's environmental control and recovery systems.*

Above: *Dr Wernher von Braun (right at rear), then Director of Development Operations Division, with other officials of the Army Ballistic Missile Agency at Huntsville, Alabama, in 1956.*

Opposite: *Dr Wernher von Braun, designer of the V-1 and V-2 rockets and now to many people the father of the US Space programme. He is pictured here beside Robert H. Goddard's space rocket.*

But we are jumping ahead in our story. The fact that the rockets which Goddard had been developing bore a strong similarity to those which von Braun had been making in Germany proved most fortuitous. In 1946, a new high-altitude rocket to supersede the *WAC-Corporal* was put on trial. Almost at once, the *Aerobee*, as it was named, proved both superior in performance and in its capacity for carrying instruments. Just over 18 feet long, it attained a velocity of 2,790 mph and could reach altitudes of 70 miles. The very threshold of space was now almost in reach.

With these new achievements, the design problems became more complex: but once again von Braun's work on the V-2's proved crucial. To breach space a perfectly controlled ascent was seen to be more important than a huge upward drive at enormous speed. From these deliberations emerged the *Viking*, a 42-foot-long rocket which achieved a thrust of

The Aerobee *rocket, 50 feet long and 15 inches in diameter, was capable of lifting a 150-pound payload to an altitude of 150 miles.*

20,500 lbs from a combination of liquid oxygen and alcohol fueling. The first of this successful series was launched from White Sands in May 1949, and from its simple launch pad the projectile broke through the magical 100 mile mark.

The development of the *Viking* continued apace into the 1950s, each new model achieving still higher altitudes. Then, in July 1955, a significant moment occurred when it was announced from the White House that the government had approved the construction and launching of a small, unmanned satellite to be put into orbit around the Earth. The initial launching, it was stated, would take place during International Geophysical Year: July 1957 to December 1958.

There is no doubt that this achievement was within the capabilities of the Americans – but behind the scenes at NACA there was inter-service rivalry between the Army and the Air Force, particularly about the use of rockets. The wrangling denied von Braun the approval to launch the satellite and then, right out of the blue, the USSR's *Sputnik* (the Russian word for satellite) was suddenly heard bleeping its way around the globe.

The Russian triumph stunned the Americans who had hardly considered they had any rivals in the race for space – and there were those in the country who were less than generous in their praise of the achievement. But attitudes did change after von Braun and his team sent up the first American satellite, *Explorer 1* on January 1 1958, and the mission also became the first space vehicle to make a major scientific discovery – that of the Van Allen radiation belt which surrounds the Earth at a height of 550 miles. This pioneer satellite which was placed in a 223 by 1,575 mile orbit remained functioning for over 12 years before burning up in the Earth's atmosphere in March 1970.

That same year also saw the resolution of the NACA problems by its re-establishment as NASA and the beginning of an era of achievements in space that grows ever more impressive year by year. Its major triumphs are familiar to everyone: twelve astronauts have now walked on the Moon, and dozens more have flown in space. The planets of Mercury, Venus, Mars, Jupiter and Saturn have been circled and photographed by unmanned probes (with *Voyager 2* now on course for Uranus and Neptune, due in January 1986 and August 1989 respectively) while the Space Shuttles are the admiration of the entire world.

But some of the Agency's smaller, less publicised achievements are also worthy of record, I think, and deserve a place in these pages.

For instance, in February 1959, the first of the *Discoverer* series of satellites was put into polar orbit, to be followed almost immediately by the hugely successful weather satellite, *Vanguard 1*, the fore-runner of a host of such vehicles (*Tiros, Nimbus, Meteosat*, etc.) which now provide a continuous watch over the whole of the Earth's atmosphere, along with the most detailed meteorological reports.

Communication satellites began with *Score* which received and transmitted actual voices: a high point of this particular pioneer's life-span being the relaying of a Christmas message from President Eisenhower, a few weeks before it decayed in January 1959. The first television satellite, *Telstar*, was launched in July 1962 in an orbit ranging from 600 to 3,500 miles, and on July 23 of that year it successfully transmitted the first pictures of an American baseball game to British TV viewers.

In the wake of these NASA satellites have come a great many more put into orbit by other nations. The first international satellite co-operation was in fact between Britain and America when *Ariel 1* was launched in 1962. The Canadian's *Alouette 1* began to orbit that same year; France's *A-1* in 1965, Japan's *Osumi* in 1970; and China's simply named *China 1* the same year. The first all-British satellite, *Ariel 3*, was launched in 1966 and now forms part of the giant, multi-national network of satellites relaying information, television, and radio signals all over the world. (Mention should also perhaps just be made of *Earlybird*, the first commercial TV satellite put into orbit in 1965.)

NASA has also been responsible for launching the navigational satellites, *Transit*, and the Earth survey vehicles called *Landsats* which report on geographical conditions, human activity and a variety of other functions for the benefit of mankind. Equally, of course, satellites have been put into orbit for military purposes. The Americans launched the first in May 1961 with *Midas 2* which carried infra-red detectors capable of giving warning of an imminent missile attack. Subsequent vehicles have included the *Big Birds*, capable of incredible picture definition, and the *Navstars* which pinpoint the movements of naval and ground forces. The Russians have responded with satellites such as the lengthy *Cosmos* series which have a variety of capabilities from orbital bombardment to interception of other spacecraft.

At the time of writing this book (Autumn 1984), space observers estimate that America has 50 military satellites in orbit as compared to the Russians' 90 – all dedicated to spying and increasingly likely to be the object of anti-satellite warfare. The dangers of such action is making ever more urgent the need for dialogue between the two super-powers about banning space weapons.

It has, though, been in the area of manned space flight that NASA has registered some of its finest achievements. Although the Russians put the first animal in space – the ill-fated puppy Laika in November 1957 – America successfully launched and returned to Earth two monkeys, Able and Baker, in May 1959 after a 15-minute flight. And one of these 'monkeynauts', Baker, has in fact survived to celebrate NASA's 25th Anniversary at his home in the Space and Rocket Museum at Huntsville, Alabama.

There was much less of a gap when it came to putting the first human being in space. Again it was a Russian, Yuri Gagarin, who made the first orbital flight on April 12, 1961, but only three weeks later on May 5, Alan Shepard followed him in a 15-minute sub-orbital flight in his tiny, bell-shaped, one-man capsule, *Mercury*. Less than a year on, the first orbital flight was achieved by John Glenn in *Friendship*. And by the time of *Mercury 9*, 22 orbits were achieved in a mission of just over one hour, and the challenge which President John Kennedy had set NASA in 1961 to put a man on the Moon by the end of the decade began to look rather more possible than problematical!

Indeed to achieve this object, NASA had already set in motion *Project Gemini* in which astronauts were being trained for the various demanding tasks which such a challenge presented: namely orbiting the Earth and then breaking out of the orbit and heading for the Moon. There was also the complicated manoeuvre of actually landing on the lunar surface and then setting off for home. The *Gemini* spacecraft were similar in appearance to the *Mercurys*, though they weighed twice as much, were far advanced technologically and carried two men instead of one. It was perhaps appropriate that as this Moon lander owed more than a little to the BIS spacecraft designed all those years earlier, a vital part of its equipment, a fuel cell, had been developed in Britain. This cell generated electrical power by

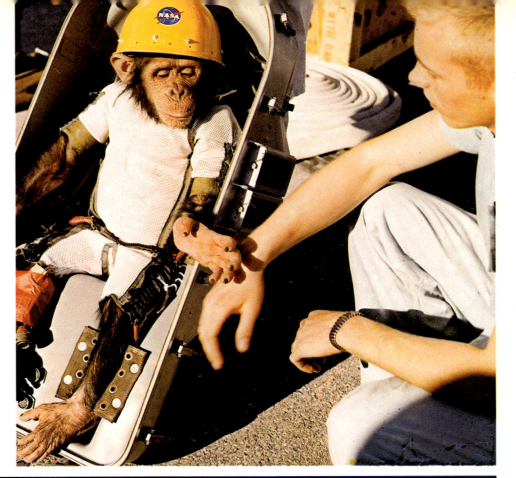

Above: *Alan Shepard in the Project Mercury spacecraft in which he made the first American sub-orbital flight.*

Top left: *One of the chimpanzees, wearing a NASA hat, specially trained for the* Mercury-Redstone 2 *flight, pictured in the chair to which it was strapped during the 16-minute ballistic flight from Cape Canaveral.*

Bottom left: *This photograph of NASA's* Gemini VII *was taken through the hatch window of* Gemini VI *during rendezvous at an altitude of approximately 160 miles on 15 December 1965.*

111

combining hydrogen and oxygen, and as a by-product of this made drinking water for the crew.

In the years 1965 and 1966, ten of these *Gemini* capsules were flown, and the necessary skills associated with long-duration flight, rendezvous with other vehicles, docking, space walks and guided re-entry were all mastered by the astronauts. The highlights of this programme were undoubtedly Edward White's space walk from *Gemini 4*, the fortnight-long, 206 orbits of *Gemini 7*, crewed by Frank Borman and James Lovell, and Neil Armstrong and David Scott's historic space docking with *Gemini 8*. The successful achievement of this procedure at last made a Moon landing feasible, and it seemed only right that the man who had performed it should have the honour of being the first man to step onto the lunar surface.

The 3-man *Apollo* spacecraft which finally achieved President Kennedy's objective – though with only three months to spare – remains to this day one of the most famous pieces of space hardware, perhaps only challenged nowadays by the versatile Space Shuttle. Based, as I said on the BIS idea, it was launched by the rocket which was the culmination of Wernher von Braun's technological skill, the mighty *Saturn 5*. But getting to the Moon was no easy matter, however.

The first actual test of an unmanned *Apollo* Command Module took place as early as May 1964. But disaster overtook what was planned as the first manned trial flight from Cape Canaveral in January 1967. An electrical fault which materialised just before take-off caused a fire in the spacecraft, and the three crew members, Gus Grisson, Roger Chaffee and Edward White, were burned to death before rescuers could reach them. This tragedy caused a total re-think by the NASA scientists. It was evident that the undue haste with which the *Apollo* system had been designed was principally responsible for the fire and so the whole thing was redesigned. Though this accident was a set-back to the programme, it brought about vital rethinking without which the Moon mission might have failed at a later and still more damaging point.

The *Apollo* triumphs which followed are so well-known as to need no more than the briefest resume here. Suffice it to say that *Apollo 8* crewed by Frank Borman, James Lovell and William Anders became the first manned spacecraft to leave the Earth's orbit in October 1968. Two months later, as the mission flew around the Moon, Borman sent an unforgettable Christmas

THE WINGS OF THE EAGLE

Day message back to Earth. Then, just seven months later, *Apollo 11* with astronauts Neil Armstrong, Edwin 'Buzz' Aldrin and Michael Collins, made the 102 hour journey to the Moon, and while Collins remained 60 miles above in orbit, Armstrong and Aldrin took the lunar module *Eagle* down onto the lunar surface. After touch-down on the Sea of Tranquillity, Armstrong emerged for his two hours and fourteen minute walk and uttered those now immortal words, 'That's one small step for a man, one giant leap for mankind.'

At that moment on July 20, 1969 a dream born in the earliest days of human history was fulfilled. And the safe return of the three men to a splashdown in the Pacific Ocean not long afterwards provided a second great moment in history.

In the years which followed another eleven *Apollo* manned flights were flown and the programme proved an outstanding success. Even what looked like becoming a disaster when an oxygen tank exploded in *Apollo 13* in April 1970 when it was 200,000 miles from earth, was turned into a triumph for human ingenuity when the three-man crew safely nursed their vehicle home. Although nine landings on the lunar surface had been originally planned to explore the surface and bring back samples, only six actually took place before the programme ended in October 1977. *Project Apollo* had cost the United States a cool 24 billion dollars, but in terms of scientific knowledge gained and human endeavour perfected it was beyond price.

To succeed the *Apollo* programme, NASA next launched, in May 1973, a Skylab Orbital Workshop (OWS) using the same *Saturn 5* Rocket. Weighing 90 tonnes, this orbiting workshop was the heaviest as well as the largest object so far put into space. Although slightly damaged during the launching, Skylab played home to three crews of astronauts who lived and worked there for periods of 28, 59, and 84 days respectively.

The 48-foot-long Skylab, sheathed in a film of gold to protect it from the solar heat, is rather like some complex factory building compressed to a fraction of its normal size. Exploring the stand-by version of the craft which is today on show in the National Aeronautical and Space Museum in Washington gave me the feeling of being in the same kind of clinical and highly organised atmosphere one finds in a hospital: functional and a bit awe-inspiring. Despite being only 22 feet in diameter, every inch of space has

Left: *Edwin 'Buzz' Aldrin pictured beside the lunar module* Eagle *during the first Moon landing, by* Apollo 11 *on 20 July 1969.*

Top: *The ability to live and work in space and carry out a multitude of functions is of prime importance for conducting manned missions to the far reaches of the universe.*

Above: *Skylab, the first American space station, was launched on 14 May 1973 and manned on three different occasions by three-man astronaut teams for a total of 171 days before returning to earth. The makeshift parasol constructed by the astronauts to protect them from the intense heat of the sun, is visible over the rear section. The original solar panels broke off during launch.*

been maximised, from the crew's living quarters (complete with loo) to the laboratory where the prime functions of studying the Earth, observing the Solar System and carrying out experiments in the weightlessness and vacuum of space take place.

The success of Skylab clearly demonstrated that man could exist in space for long periods, and carry out work effectively in the unusual conditions which exist there, and pointed the way ahead to what could be achieved. It is only to be regretted that the importance of the OWS tends to be overshadowed in many people's minds by the *Apollo* missions, and the unmanned space probes, *Viking, Voyager* and *Pioneer* which have more recently gone exploring the planets of our system, and sent back the information and photographs which make up a later section in this book. For it seems beyond dispute to me that the lessons learned from Skylab provided a big step forward in one of the greatest of all space objectives: to conduct manned missions to the far reaches of the universe, when the ability to survive long periods in space and carry out a multitude of functions will be of prime importance.

The latest of NASA's space vehicles, the Space Shuttle, has taken this potential further, for it will be able to assist in the building of the Space Stations from which interplanetary missions will surely begin. The remarkable orbiter which has made flying from Earth into space and back again such a comparatively easy matter, actually came about as a result of military and financial pressures which were put on NASA to find a compromise between the one-shot rocket and a spacecraft that could be used over and over again.

Initially, the Shuttle evolved from the secret X-15 rocket aircraft of the late 1950s by way of three experimental craft, the M2-F2, HL-10 and X-24, developed and tested by NASA at the Langley and Ames Flight Research Centers. The key to success came when the original rather oval shape of the craft was dramatically thrown out for the present delta form complete with its blunt undersurface to absorb the atmospheric heating in flight and permit a gradual descent and smooth landing. To power the craft liquid fuel engines were designed, said to be capable of generating the same power as 23 Hoover Dams!

The first of these 'Lifting Body' Shuttles, *Enterprise*, was flown in the

Earth's atmosphere in 1977, piloted by *Apollo 13* veteran, Fred Haise. This craft never went into orbit, and it was not until three years later – and in the teeth of persistent rumours that NASA was having untold problems with its experimental orbiter – that *Columbia* at last became the first Shuttle in space in 1981, commanded by John Young. Since then, two further orbiters have come into service: *Challenger*, which on October 11 1984 provided astronaut Kathy Sullivan with the opportunity to become the first American woman to walk in space some 137 miles above the South Atlantic; and *Discovery*, which made its maiden flight in September 1984. A further Shuttle, *Atlantis*, flies for the first time in 1985.

And so the projects go on at NASA's various centres: from the headquarters in Washington to the far-flung Jet Propulsion Laboratory in Pasadena, California where deep space programmes are under active preparation; at Ames Research Center in Mountain View, California, looking at the origins of life; at the test flight centres of Hugh L. Dryden Flight Research Facility at Edwards, California; and the Robert H. Goddard Space Flight Center in Greenbelt, Maryland. Not forgetting the George C. Marshall Center at Huntsville, Alabama, which tests engines, the Langley Research Center in Hampton, Virginia, developing advanced flight concepts, and the Lewis Research Center in Cleveland, Ohio, busy with propulsion systems. Plus the National Space Technology Laboratories in Bay St Louis, Missouri, for static firing, and the Wallops Flight Facility on Wallops Island, Virginia, responsible for flight planning and experimentation. At the heart of the empire lie the superb John F. Kennedy Space Centre in Florida where the space vehicles are launched and the Lyndon B. Johnson complex in Houston which controls these scientific marvels.

Each space achievement that NASA chalks up is due to the work of all these teams. And with 25 years of accomplishment behind them – here's to many more!

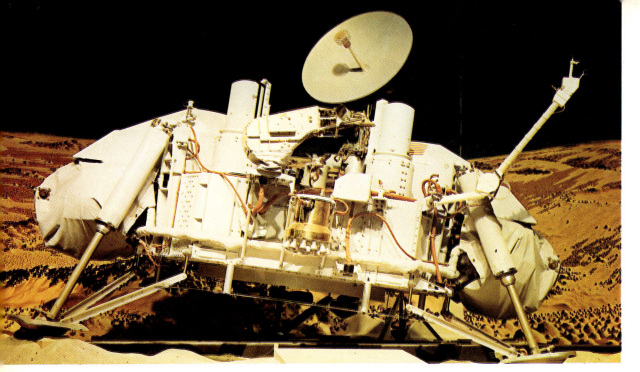

Top: *The* Viking *Test Launcher at the Jet Propulsion Laboratory in Pasadena, California, set up here to correspond with the tilted position of the genuine* Viking 2 *as it stands on the surface of Mars.*

Above: *The Jet Propulsion Laboratory's Deep Space Network, in Pasadena, California, communicates with, and tracks automated scientific spacecraft travelling in deep space. Pictured here is the JPL's control centre.*

Opposite: *The Space Shuttle* Challenger, *taking off from Kennedy Space Center, Florida, on its maiden flight in April 1983.*

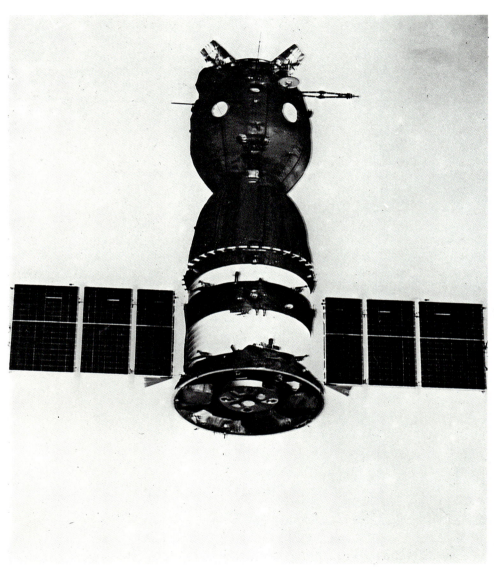

The Soviet Soyuz *spacecraft in Earth orbit, photographed from the American* Apollo *spacecraft during the joint US-USSR* Apollo-Soyuz Test Project *docking mission.*

120

THE
SPUTNIK LEGACY

THE COMMITMENT of the Russians to the achievement of space travel has been as long standing and in quite a number of respects as successful as that of the Americans, and consequently no book such as this could hope to be considered balanced without some study of their record. That their space work is not better known is almost entirely due to the veil of secrecy which they draw over their experiments – and, consequently, piecing together what information there is proves a painstaking task.

In Russia, as elsewhere, interest in rocketry was evident over many centuries, and indeed according to one account a 'Rocket Research Establishment' was founded as early as 1680. The purpose of this establishment was undoubtedly directed towards the use of rockets in warfare, and it is not until the end of the nineteenth century that any kind of proposal for a rocket 'flying machine' was made.

Such an idea was put forward in 1881 by a remarkable man named Nikolai Kibalchich (1854–1881), an explosives expert, who had made the bomb used by a group of revolutionaries to kill Tsar Alexander II in March of that year. While awaiting sentence of death, he devised a heavier-than-air machine with a reaction motor. This craft rather resembled a flying platform with a cylindrical chamber mounted on top.

According to Kibalchich, charges of compressed gunpowder were to be fed into this chamber, and exploded so that the force was expelled through a hole in the base of the platform. This would cause the machine to rise vertically until it had reached a suitable altitude at which point, said Kibalchich, it could be swung into a horizontal position for the rest of the flight.

The velocity of this 'rocket platform' would apparently be controlled either by the size of the cartridges or the rapidity with which they were fed into the explosion chamber. However, the young revolutionary was executed before he could develop his plans any further, and in fact all details of his extraordinary design were unknown until after the Russian Revolution.

Two years after Kibalchich's execution, however, another Russian began to work quite independently on the idea of rocket propulsion, and as a result of his achievements he has become regarded in many scientific circles as the first man to fully appreciate the true significance of rockets for use in space travel. The man's name was Konstantin Tsiolkovsky (1857–1935).

'For a long time I thought of the rocket as everybody else did,' Tsiolkovsky was to say years later, 'just a means of diversion and of petty everyday uses. I do not remember what prompted me to make calculations of its motions. Probably the first seeds of the idea were sown by that great, fantastic author, Jules Verne – he directed my thoughts along certain channels, then came a desire, and after that, the work of the mind.'

The idea of interplanetary travel seized Tsiolkovsky's mind when he was still in his teens, and it was not long before he had come to appreciate the limitations of solid fuels like gunpowder as a means of propulsion. Building small rocket-shaped models he developed a theory utilising liquid fuel: one liquid being the fuel and the other – the oxidant – causing it to burn. Liquid oxygen and liquid hydrogen seemed the ideal pair to him, and he came to the conclusion that the expansion of these gases in a combustion chamber would produce a hot exhaust strong enough to propel a vehicle.

Tsiolkovsky revealed his findings in a remarkable article entitled 'The Exploration of Space with Reactive Devices' which he published in the scientific journal, *Nauchnoye Obozreniye*, in 1903. Through the use of liquid fuel, he wrote, he had come to the conclusion that it would be possible for a suitably designed rocket ship to explore the upper atmosphere of the Earth, and even contemplate interplanetary flight. To support his contention, a sketch of the kind of ship he had in mind was included (see diagram).

Nor was this Tsiolkovsky's only proposition. Later, he put forward the idea of a multi-stage rocket with each part being jettisoned once its fuel was used up until a final passenger-carrying capsule travelling at high speed was freed to explore the universe. The remarkable Russian also correctly predicted the zero gravity situation which would be experienced in space flight, and suggested the use of green plants to remove excess carbon dioxide in long space flights. He even looked forward to the day when men could build orbiting space stations and explore the surfaces of other planets.

'Mankind will not stay on the Earth forever,' he wrote towards the end of his life, 'but, in the pursuit of the world and space, will at first timidly penetrate beyond the limits of the atmosphere and will then conquer all the space around the Sun.'

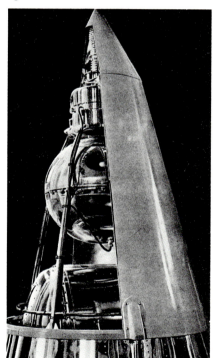 

Left: Sputnik 2 – *The lower part is the container which carried the dog, and above this are further containers for scientific equipment.*

Right: *Major Yuri Gagarin of the Soviet Air Force, the first man to make an orbital flight around the Earth.*

Strikingly prophetic though much of what Tsiolkovsky wrote proved to be, he never went any further than outlining his ideas – and it was left to more practically-minded people to make them reality, such as Robert H. Goddard in America developing liquid fuels in the twenties and thirties, and the British Interplanetary Society members beginning the evolution of the space ship.

It was undoubtedly the general political and economic instability in Russia during the first decades of this century that were important factors in preventing the country taking up the lead which Tsiolkovsky might well have given them. There were, though, rocket enthusiasts just like those in America and Britain who tried to push back the space frontiers.

Inspired by Tsiolkovsky's writings, groups of people in Moscow, Leningrad and several other of the larger cities, set up loosely-knit groups to conduct speculative research and stage small-scale rocket trials. A leading figure among these enthusiasts was a certain Valentin Glushko who, like Wernher von Braun, had become smitten with space travel in his youth and by the mid-1920s was hard at work on his own ideas. Around 1929 he built his 'Helio Rocket Plane' which actually worked by electric power. This spaceship idea consisted of a hollow sphere with a series of electrical rocket engines mounted around the circumference, the energy being fed to them from a mirror trapping the Sun's rays!

Though this idea was basically impracticable, it did provide Glushko with the stepping stone to an eminently more workable liquid fuel rocket engine, the *Optynyy Raketnyy Motor*, or ORM, which he completed in 1931. This, the first Russian liquid rocket engine, produced 44 lbs of thrust when burning a mixture of liquid oxygen and gasoline. It performed most satisfactorily when first demonstrated in Leningrad at the end of the year.

Another important Russian rocketeer was Sergei Korolev, a member of one of the country's earliest rocket societies, the *Gruppa po Izucheniyu Reaktivnogo Dvizhenia* (Group for the Study of Reaction Motion), known as GIRD for short. Korolev, an engineer by training, was destined to become one of the most famous of the early Soviet rocket designers, but initially was more taken with the sport of gliding and hankered after designing racing aeroplanes. However, he became fascinated with the idea of space travel after discovering the articles written by Tsiolkovsky, learning about Robert Goddard's work in America, and also seeing the remarkable German film,

Die Frau Im Mond (The Woman in the Moon) released in 1928 and on which Professor Hermann Oberth had acted as technical adviser.

From his own research, Korolev became convinced that the most likely way of solving the problem of space flight was by combining a rocket engine with a type of aeroplane design. For his engine he took Glushko's ORM and incorporated this into his own concept of a 'flying wing'. Although the Raketoplan (Rocket Plane) which he later unveiled made nine flights in all – some of them with the motor installed – it never actually flew under its own power.

In 1933, however, a variation of this plane, more highly streamlined and with the rocket motor centrally-located, did leave the ground for a short flight. The success of the GIRD 09, as it was called, not only excited much public interest, but was also undoubtedly instrumental in the Russian authorities' decision to set up the *Reaktivni Nauchno-Issledovatel'kii Institut* (Scientific Research Institute of Jet Propulsion, or *RNII*) which has been described by the Soviets as the first state rocket research establishment in the world. This body brought all the various groups of enthusiasts and scientists under one state-controlled umbrella. The only person who would not become involved with the *RNII* was the great Tsiolkovsky, who was now well over 70 and preferred to live and work in quiet isolation.

It is a surprising fact that despite these achievements, Soviet interest in space travel all but disappeared in the years just prior to the outbreak of the Second World War. The reason for this seems painfully simple: Stalin's infamous purges of many levels of Soviet life – in particular scientific circles. Though much of this period remains shrouded in mystery, there is no denying several leading rocketeers died in strange circumstances and much of their research was halted. It is a fact, too, that the only practical result of all the rocket research was the development of a single solid-fuel barrage weapon called the *Katyusha* used by Russian soldiers in the war.

Miraculously, though, both Valentin Glushko and Sergei Korolev survived the purges and later re-emerged to become the leading rocket builders of the *Sputnik* era.

Like the Americans, the Russians were enabled to make a quantum leap on the road towards the conquest of space after the war by the capture of some of the German V-2 rockets and several of the scientists and technicians who had worked on them. Although the Russian in charge of supervising these men, G. A. Tokaty, later claimed that his own people were as 'advanced, inventive and clever as the German rocketeers', he had to concede that in putting the theories into practice, 'we appeared to be miles behind.'

Once, however, they *had* absorbed these men and materials into their programme, the Russians worked quietly and unobtrusively until that dramatic day in October 1957 when they launched the *Sputnik* around an amazed world.

It was, in fact, with a modified ballistic missile that the USSR effectively opened the 'Space Age' with the launching of the world's first artificial satellite. The missile carried the rather unprepossessing football-sized sphere on its nose cone, jettisoning it into an elliptical orbit at an altitude of between 141 and 588 miles. Thereafter for 21 days it circled the globe transmitting its memorable 'Bleep. Bleep. Bleep.' signals back to Earth. Sputnik remained in orbit until January 1958 when it finally decayed in the atmosphere after completing 1,400 orbits.

While the Americans – who could scarcely believe they had a serious rival in space – were still catching their breath, the Russians followed up this achievement in November 1957 with a second *Sputnik* which carried a small dog named Laika and some equipment to carry out geographical and biomedical experiments. The fact that there was no hope of retrieving the little animal somewhat muted the event. Two years after this, *Luna 1* became the first solar orbitting satellite.

Though the Americans speedily began to match the Soviet Union in launching satellites, it was still the Russians who succeeded in putting the first man in space on April 12, 1961. (A year earlier, incidentally, they had successfully launched a rocket carrying a dummy man in May 1960.)

The first man to make an orbital flight around the Earth was, of course, Major Yuri Gagarin of the Soviet Air Force, who remained in space for just one hour 48 minutes to achieve a single circuit in his tiny craft, *Vostok 1*. The unassuming Gagarin merely said of his outstanding achievement, 'I am still an ordinary mortal, and have not changed in any way.'

He may not have changed – but mankind's perception of space certainly had. It was now quite clearly conquerable by him.

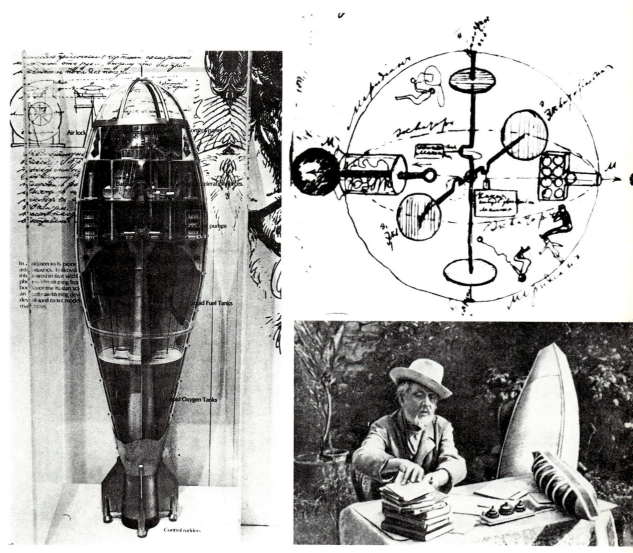

Left: *This tear-drop shaped spaceship was designed by Tsiolkovsky in the early twentieth century.*

Top right: *Sketch of cross-section of Tsiolkovsky's jet-propelled spaceship.*

Above right: *Konstantin Tsiolkovsky, working on the design of his all-metal airship, 1933.*

The success of the mission was greeted with unstinted praise by people like Sir Bernard Lovel of Britain's Jodrell Bank Telescope who said, 'This is the greatest scientific achievement in the history of man. The next step will be to keep men in space for one or two days. After that they will orbit the Moon and eventually land on it.'

Like a number of other experts, Sir Bernard was quick to point out that those who had been sceptical about manned space flight – in particular what they believed to be the dangers which would make it impossible – had been proved wrong on almost every count. Gagarin had not been space-sick, nor had he been seared by cosmic radiation or battered by pieces of meteorites. He had also coped with the state of weightlessness without any problems.

The first Russian cosmonaut was soon followed by the first American astronaut, John Glenn, in 1962, but the Russians had still one more ace up their sleeve before having to concede some of their lead to the Americans. On June 16, 1963, Valentina Tereshkova became the first women in space.

Though, as I said earlier, the Russians release very little information about their space programme, and have a tendency to announce their achievements only after the events have occurred and thereby generate a certain amount of understandable scepticism, it is still possible to assemble a general picture of their activities.

Only the most cursory glance at the figures reveals that the Soviet programme is characterised by a high level of space launches – averaging about 80 lift-offs every year since 1969. (A figure more than double, and occasionally three times, that of the USA.)

Although the programme is under the direction of the Central Committee of the Communist Party, several subsidiary organisations actually carry it out. The cosmonauts are trained by the Air Force, which also conducts research into space physiology. The Strategic Missile Force supervises the launchings of all rockets and their payloads, while the Ministry of Aviation Industry is responsible for producing the hardware. All military aspects of the space programme are, naturally enough, sponsored by the Ministry of Defence.

The main training centre for the Soviet cosmonauts is at a location called Star City just outside Moscow, while the space vehicles in which they undertake their missions are launched from three ranges. The first of these is the Baikonur Cosmodrome in Kazakhstan which is used for all the *Soyuz* flights and has more than a little similarity to the Kennedy Space Centre at Cape Canaveral. The second centre is at Kapustin Yar on the lower Volga River, which is generally viewed as the opposite number to NASA's Wallops Island centre in Virginia. The third location is the busy Plesetsk Base near Archangel which photographs have revealed to look like the Vandenberg complex in California.

The main control centre for both manned and unmanned space flights is situated in the Crimea, though a new manned flight control centre has just recently been established at Kaliningrad near Moscow. There are also spacecraft tracking stations throughout the USSR, with the network extended through bases in other Warsaw Pact nations as well as tracking ships at sea.

As to the programme itself, after the initial two *Sputnik* launches, the Russians next made six flights in four of the heat-shielded *Vostok* spacecraft in the years between 1961 and 1963.

Then, on October 12, 1964, an improved version of the *Vostok*, called a *Voskhod*, was launched with three cosmonauts squeezed into the tiny interior. These cramped conditions and the fact the men were not wearing space suits has led to suggestions that the flight was hastily ordered by Premier Krushchev in order to claim the distinction of the first group of three men in space before the American *Apollo* flew. A second, and final, *Voskhod* was launched in 1965 and this mission was highlighted by the first demonstration of 'Human Extravehicular Activity' (EVA), the space walk, when Lieutenant Colonel Alexei Leonov left the craft and was outside the capsule for almost 20 minutes.

Two years later, in 1967, the Russians launched the first of their now highly successful, multi-purpose *Soyuz* craft which have averaged more than two flights a year ever since. These vehicles, which can be used as a small space station, as a rescue and recovery craft, and as a ferry service, consist of a three-manned command and re-entry module, a service module, a work compartment and a solar energy source.

Despite its later success, the *Soyuz* was nevertheless beset with problems during its early days. The re-entry module proved too small for three cosmonauts in their bulky spacesuits, and there were seemingly endless

This photograph of the Soyuz *spacecraft, taken from its American counterpart, shows its three major components – the spherical-shaped Orbital Module, the bell-shaped Descent Vehicle, and the cylindrical-shaped Instrument Assembly Module.*

Opposite: *An artist's impression of the* Apollo-Soyuz *Test Project (ASTP) docking, July 1975.*

manoeuvring troubles. *Soyuz 1* was also responsible for the first space disaster in 1967 when Cosmonaut Komarov was killed, and tragedy also struck *Soyuz 11* in June 1971 when all three crew members died as a result of pressurisation failure as they returned from a 24-day flight.

Gradually, though, these problems were overcome. *Soyuz 19* was able to successfully rendezvous with the NASA *Apollo* spacecraft for 47 hours on July 17 to 19 1975, astronauts Tom Stafford, Donald Slayton and Vance Brand performing the first space handshakes with the Russian cosmonauts Alexei Leonov and Valeri Kubasov, to the delight of the millions watching television below on Earth. On November 4 that same year, *Soyuz 20*, an unmanned launch, docked with a *Salyut 4* space station and carried out a variety of biological experiments – demonstrating a heretofore unknown Russian ability of being able to resupply an orbiting space station.

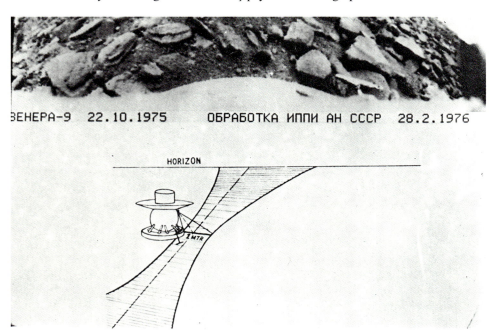

ВЕНЕРА-9   22.10.1975   ОБРАБОТКА ИППИ АН СССР   28.2.1976

HORIZON

1 MTR

A picture of the surface of Venus taken by the Soviet Union's unmanned Venus probe, Venyer 9 *spacecraft, on 22 October 1975.*

THE SPUTNIK LEGACY

This was followed in January 1978 by another historic moment with the double docking of *Soyuz 27* complete with a two man crew, to the *Salyut 6* space station to which *Salyut 26* was already attached. A remarkable three craft link-up. As a result of this, it was now possible for Russian space crews to visit the space station and, if need be, change into another vehicle.

With each passing year, the Russians have pushed further the lengths of time which cosmonauts have been able to stay in space, thanks to the reliability of the *Soyuz* craft and the capabilities of the *Salyut* space stations after a number of early problems. The first of these stations was launched as long ago as 1971, and the Soviets' continued development of their facilities shows them to be an important element in what is believed to be one of their prime objectives in space – a network of permanently manned orbiting stations. At the time of writing, there have been seven *Salyuts* in orbit, the most recent of these having been 'home' to cosmonauts Leonid Kizim, Vladimir Soloyov and Dr Oleg Atkov who set an endurance record of 212 days from February 8 to October 2 1984. During the men's stay in space they undertook six space walks and were twice visited by *Soyuz* teams – including the second Russian female cosmonaut, Svetlana Savitskaya, who became the first woman to walk in space in July 1984.

In essence, the *Salyut* is a 15 metre-long set of cylinders, made up of a 4.15 metre diameter workshop, a narrower 2.9 metre control station, and a two metre wide airlock at the nose. Attached to the mid sections of the later versions are three large solar panels fixed at 90 degrees from one another. These models also have docking facilities at both ends.

The purposes of the stations have varied between civilian application and military uses, although there has been much general experimentation with solar studies, geophysical phenomena, biology and Earth resources. Supplying the *Salyuts* with fuel and food for the cosmonauts has recently been undertaken by an unmanned vesion of the *Soyuz* vehicle known as a *Progress* which flies up from earth. This craft can also act as a 'Space Tug' to push the *Salyut* back into orbit every time it begins to slip back towards Earth. Although the *Salyuts* weigh probably less than one third of the American *Skylab* station, they are generally considered to be rather more versatile.

The Russians have continued the lead of *Sputnik* with a very active programme of satellite launches. The *Cosmos* series which began in 1962 represents the largest number of Soviet space vehicles orbited so far – over 2,000 in fact. They are used in a variety of roles including space environment observation, Earth reconnaissance, biomedical experimentation, meteorological studies, navigational aids, and in a variety of military related operations. Indeed, it is fair to say that the majority of Russian satellite launches are concerned with photographing foreign military installations, ocean surveillance, and the study of means for utilising these spacecraft as part of weapons systems.

The first successful Russian communications satellite was the *Molniya 1* launched in 1965, and since that date over 200 more have been put into orbit. They are linked to a huge network of Orbita ground stations with the Soviet Union and ships in the Atlantic and Pacific Ocean which receive and transmit a busy range of telephone, telegraph and television traffic. There are also a group of scientific satellites bearing the various designations *Polyot*, *Electron*, *Proton*, *Prognoz* and *Intercosmos*, which have been involved in investigations of the Earth's radiation belts and magnetic fields, cosmic rays, the ionosphere, and a variety of other near and deep space environmental phenomena.

Although due to the success of the *Apollo* programme, the Americans have grabbed the bigger share of headlines where the Moon is concerned, the Russians have also been busy for years studying our nearest neighbour. *Luna 1*, launched in 1959, was in fact the first satellite to pass near the Moon, while *Luna 2* hit it, and *Luna 3* photographed the far side. All of these satellites only provided the crudest photography until *Luna 9* made the first survivable landing in 1966. Then from 1969 there were larger payload flights, *Luna 16* returning with soil samples from the lunar surface, and *Luna 17* and *Luna 21* putting the extraordinary looking Lunokhod vehicles down to roam the barren wastes for several months and relay information back to Earth.

As far as deep space is concerned, the Russians have launched almost 50 satellites in the years since 1960, though the results they have achieved from these probes have been disappointing. They have been rather more successful with the operating payloads known as *Veneras* which they have managed to put on both Mars and Venus – as well as orbiting satellites around these planets – but specific information regarding their findings has been

East meets West in space. Astronaut Thomas P. Stafford and Cosmonaut Alexei A. Leonov meet in the docking hatchway between their two spacecraft on 30 July 1975. The Apollo *and* Soyuz *spacecraft remained docked in Earth orbit for two days.*

sparse and less detailed than that made known by the Americans.

Perhaps not surprisingly, it has come to light recently that the Russians are hard at work trying to develop a Space Shuttle to match the American's successful vehicle – and the first indications are that the Soviets are copying both the design and technology of their rival. Although smaller in size (109 foot long with a 76 foot wingspan), the *Raketoplan* – named after Glushko's pioneer craft – has one unusual feature: a highly aerodynamic pointed tail cone with several small fins. It is expected that the spacecraft will make its initial manned trials from the back of a Myasishchev Mya-4 'Bison' bomber aircraft (just as NASA's first orbiter, *Enterprise*, did from the back of a Boeing 747 in 1977), but it is most likely to be launched into space on top of the D-Type rockets used to send up the *Salyut* space stations and planetary probes.

Soviet sub-scale shuttle being recovered from the sea after a test flight. (photo: Dept. of Defence, PRO Canberra)

According to the most recent information, the first tests of a sub-scale shuttle were conducted at the Kapustin Yar launch site in June 1984 with the model being recovered later from the Indian Ocean where it parachuted down at the end of the flight. This was followed by approach and landing tests at the Ramenskoye Flight Test Centre near Moscow, and most recently – on December 21, 1984 – it was said that a one-third size shuttle had been successfully launched into orbit and made a 'controlled descent' back to Earth.

It now no longer seems a matter of *whether* the Russians will have their own manned reusable space orbiter, but *when*. And just as this vehicle has opened up whole new horizons in space for NASA, so it will surely do for the Soviets.

THE EUROPEAN CONNECTION

THE 'THIRD POWER' in space is undoubtedly the multi-national union of European nations known as the European Space Agency, or ESA for short. The group consists of Great Britain, Belgium, Denmark, France, West Germany, Italy, Holland, Spain and Switzerland, with Austria and Norway as associate members.

The ESA came into being in 1973 when eleven European countries, realising they could never afford space programmes such as the Americans and Russians, decided to pool their resources, each partner contributing a proportionate amount. The size of Britain's contribution places them in the middle of the group.

The purpose of the Agency was clearly laid out in its convention which said it was to 'provide for and promote, for exclusively peaceful purposes, co-operation among European States in space research and technology.' It was set up, in fact, on similar lines to NASA and it was decided to make its headquarters in Paris.

Because, not surprisingly, there are no suitable launching sites for large rockets in Europe, ESA has had to develop its own launch centre amidst the lush jungle landscape of Kourou in French Guiana, a locality which has caused it to be described as a mixture of 'science city and Caribbean holiday-style living.' (Probably the only other suitable site would have been the old

Opposite: Ariane *lifts off from the ESA's space centre at Kourou, French Guiana.* Ariane *has come to rival the Space Shuttle both in terms of capability and cost. Its major drawback, however, is that it is not reusable.*

139

Woomera Base in Australia where Britain had tested rockets back in the early 1960s.)

Despite its obvious lack of vast financial aid, the Agency has become increasingly successful during the past decade. It now builds satellites, manufactures and flies rockets, and has developed *Spacelab*, the multi-purpose unit for studying the Earth's atmosphere and conducting experiments in space physics.

In its early days, ESA had to rely wholly on the Americans to launch its space vehicles, but since 1979 it has had its own rocket launcher, the *Ariane*, and consequently its programme has become increasingly independent. Indeed, such has been the success of this rocket, that ESA is beginning to compete with NASA in offering to other 'space minded' nations the facility of placing satellites in orbit, and to date has won contracts from Australia, Canada, Brazil, China, the Arab League and even from six American telecommuncations companies!

The white, snub-nosed *Ariane* which has been described as the most advanced conventional satellite launcher in history, is a three-stage, liquid fuelled rocket, which has grown in height and power since the original design was first flown in 1979. The first stage of the rocket, the L40, has four Viking 5 engines, while the second, the L33, has a single Viking 4. The third stage, like the NASA Shuttle's main engines, is powered by the cryogenic fuels of liquid hydrogen and liquid oyxgen – and is the first such engine to have been developed in Europe.

Ariane 2 which flew a year later is basically the same as the first model but does have a larger third stage. Because of the success of both of these, ESA felt confident enough to begin carrying commercial payloads on *Ariane 3* without test flights. This model was identical to *Ariane 2*, but, with the addition of two solid fuel boosters, providing 150 tonnes of extra thrust, strapped on to its sides, was much more powerful and could carry half as much weight again as the original *Ariane*. (It quickly proved this by launching *two* satellites into orbit with a single flight.)

The most powerful version, the *Ariane 4*, has increased thrust in all its stages, and in one version – designated the AR 42P – it will fly aided by two solid fuel boosters; while the even more sophisticated AR 44L will have no less than four liquid-fuelled boosters. Already, though, an *Ariane 5* is at the planning stage, and this will be very different again. Powered by an H60 cryogenic engine, it will be divided into stages as before, but each of these will be programmed to return to base: making it the world's first reusable rocket. Still further ahead, the technicians working on this project would like *Ariane 5* to be used to launch a small manned spacecraft called *Hermes* which could be used for ferrying men and equipment to space facilities such as *Spacelab* or an actual space station.

Not surprisingly, *Ariane* is seen as a direct rival to the Space Shuttle both in terms of capability and cost. It can release a satellite directly into a geosynchronous orbit 22,300 miles above the Earth, while the Orbiter has to manoeuvre its payload into a much more restricted orbit between 150 and 700 miles. And as experience has proved, the higher the orbit, the less the satellite is subjected to wear and the longer it will last. *Ariane* can also carry heavier loads. Conversely, of course, the rocket is not reusable at present, does not have the flexibility or crew capabilities of the Shuttle, and certainly could not rescue an errant satellite as the *Discovery* did in November 1984.

However, the only really sad fact about the success of the *Ariane* programme is from the British point of view. For the rocket is primarily French built by Arianespace on behalf of ESA – and yet was directly inspired by the superb British *Blue Streak* launch rocket which never failed in flight, but was still scrapped by the government in the early 1970s. Indeed, the only saving grace is that several items on the rocket are British-made, including the auto-pilot electronics.

To this still further example of Britain throwing away a potential lead in space technology, can be added the axed *Black Knight* and *Black Arrow* rockets as well as the successful *Ariel* series of satellites which were halted in 1982. (Only the British Aerospace *Skylark* series has survived, and after more than 400 launches in seven different countries while carrying out scientific experiments, they are now being employed in helping to improve the methods and technologies used aboard *Spacelab*.) And just to add insult to injury, remnants of the *Blue Streak* can be seen rotting not far from the *Ariane* launch site in Kourou, and caused one indignant journalist to refer to them as 'potent, or perhaps more accurately, impotent symbols of Britain's space endeavours!'

Britain has, however, played an important part in the successful

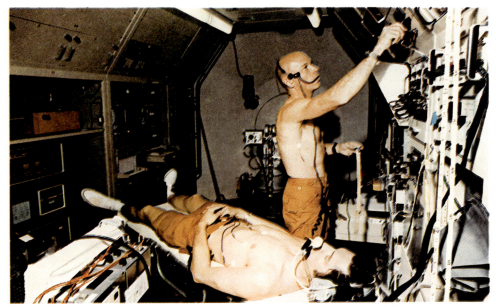

Above: *Medical and Scientific experiments being carried out in the Spacelab simulator at the Johnson Space Center, Houston, Texas, prior to launching of the ESA's Spacelab.*

Opposite: *Artist's impression of Spacelab being launched from the payload bay of the Shuttle.*

development of *Spacelab*, which many observers see as the first step towards the permanently manned space station. The orbiting laboratory is at present inextricably linked to the NASA Space Shuttle which carries it aloft to conduct a whole range of different activities.

It is probably true to say that on the day the Shuttle *Columbia* launched *Spacelab 1* – (November 28 1983) the ESA finally joined the 'big league' of nations in space. It also marked the culmination of a project which had begun well over a decade earlier when NASA had invited the European nations to participate in a joint space venture.

In 1969, in the aftermath of the *Apollo* programme, NASA set its sights on the creation of a reusable transportation system. Such a craft, with the facility to go into space and return, would open up whole new areas of human-

directed scientific and technological possibilities. The Americans decided to offer their friends across the Atlantic a chance to be involved in the project.

To look into the feasibility of co-operation, a group called the European Space Research Organisation (ESRO) was set up, and by 1972 had come up with an outline plan for a laboratory that could be flown in the cargo hold of the orbiting vehicle. A year later, when the ESA was established, the plan for *Spacelab* – as it had now become known – was given immediate priority.

The division of responsibilities was basically that NASA would build and operate the Shuttle, while the Europeans would build *Spacelab* and all the equipment associated with it. The two organisations agreed specifications for the hardware and a timetable for development and manufacture. Throughout the 1970s, work on both sides of the Atlantic quietly gained momentum until the dream became reality.

And when *Spacelab 1* was launched in *Columbia* from the Kennedy Space Centre, NASA Administrator, James M. Beggs paid this tribute, '*Spacelab* represents a major investment in the order of one billion dollars from our European friends. But its completion marks something equally important: the commitment of a dogged, dedicated and talented team drawn from ESA governments, universities and industries who stuck with it for a decade and saw the project through.'

Apart from being a milestone in space history, the launch was an important moment in the successfully developing Shuttle programme. And the six-man team also represented the largest crew ever to venture into space aboard a single spacecraft.

Spacelab is modular in design and fits into the Shuttle's 60 foot long by 15 foot wide cargo bay. The two major modules are a 23-foot long pressurised unit in which the scientists can work in a 'shirt-sleeve' environment, and an open, u-shaped pallet 13 foot wide by 10 foot long on which can be placed antennas, telescopes and other sensors to carry out experiments when exposed in space. This pallet may well consist of more than one unit depending on the number of experiments to be carried out. An 18-foot-long enclosed passageway links these modules to the Shuttle's mid-deck. Just like the Shuttle, *Spacelab* is reusable.

Britain's share in *Spacelab* has amounted to 6.3 per cent, or about £50 million per year, as against West Germany's massive 55 per cent. However, the vital experimental pallets are wholly British-made by British Aerospace and their versatility has turned them into virtual scientific observatories.

The success of *Spacelab 1* was particularly satisfying for Derek Mullinger, the British head of the team that integrated all the experiments. 'The mission generated more data than the scientists believed possible,' he said afterwards. 'And one of the greatest benefits was the ability of the specialists on board to modify experiments in mid-course and to do previously unplanned experiments as the results of earlier work became known. It was an extraordinary success.'

The outcome of the first mission was the immediate scheduling of more flights, and as many as 100 *Spacelab* missions are being talked of in the next decade. A NASA spokesman was just as enthusiastic about the project as Derek Mullinger had been.

'I have never seen a system as accurate, complete and problem free,' he said of the unique venture. 'Now it's not a case of whether we will have a space station, but when!'

ESA has, in fact, already been offered the chance of participation in the projected $8 billion station and a total contribution from the member nations of about £1 billion is being sought. Of this, Britain's share would probably be about £150 million.

Britain's involvement with ESA has naturally enough generated contracts for a number of our leading companies working in the space industry including Marconi, Ferranti, Plessey and in particular British Aerospace, perhaps the world's leading manufacturer of satellites. They, in turn, are thinking ahead with new projects, and BAe have even come up with outline proposals for a space platform and a radically new type of space shuttle! I should like to mention them here, though both are little more than ideas at the moment.

The space platform would be a free-flying module in polar orbit. It would be able to carry out such tasks as Earth observation or industrial work in the low-gravity, high-vacuum environment of space. The cost of such a project would obviously be enormous, but British Aerospace believe it is possible with ESA and NASA participation.

In particular BAe would hope to be able to draw on NASA's expertise in this area of space technology and so avoid incurring the massive development

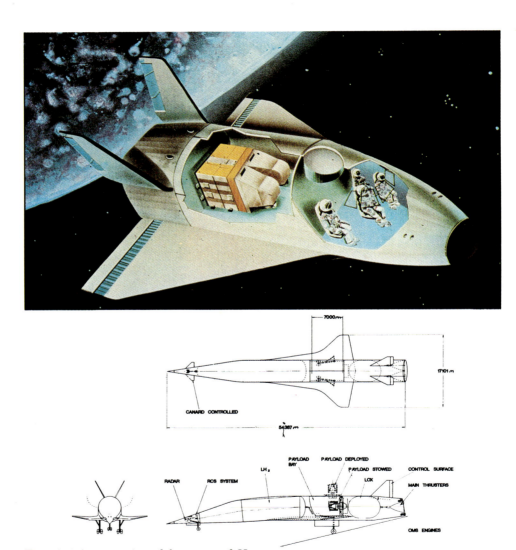

Top: *Artist's conception of the spacecraft* Hermes.

Above: *HOTOL – Concept for an unmanned Horizontal Take-Off and Landing launch vehicle, conceived by BAe and Rolls Royce to lift payloads of up to 7 tonnes to low earth orbit and to return, using standard aircraft runways.*

costs of getting such a system into space. They would hope, too, that the platform could be serviced by NASA's proposed space station.

The space shuttle idea is perhaps more of a dream – though in the light of the ill-fated Project MUSTARD it would be nice to think it *might* ultimately become a reality. The vehicle is described by BAe as a Horizontal Take-Off and Landing craft (HOTOL for short), but is completely different in both appearance and operation to the American and (embryo) Russian Shuttles.

The preliminary designs show a craft resembling a rocket rather than a shuttle. It has a needle-point nose with a pair of stabilising winglets, a pair of Concorde-type wings on the mid-fuselage, and a twin fan assembly at the rear. It also differs from the other shuttles in that it will be unmanned – equipped with an automatic piloting system to control the launch, ascent, orbital manoeuvring, payload management and return to base. It will also go straight up and into orbit rather than separating in stages: rising from its launch pad like a rocket and making a powered descent to land horizontally.

Because it carries no crew and consequently will need no life support systems, HOTOL will be significantly smaller than the existing shuttles – about 55 metres in length with the central wingspan about 17 metres. The payload bay, located just above these wings, will measure 9 metres by 4.5 metres and should be able to carry cargoes (satellites included) to a total weight of 7 tonnes.

A wholly new propellant system called the Liquid Air Cycle Engine has been devised. This is a combination of rocket and air breathing propulsion systems which enables it to carry on board all the fuel it requires for a mission. As the craft rises through the atmosphere it will draw in oxygen from the air, liquify this with hydrogen, and then utilise the mixture in its rocket engine. This system does away with the need to carry heavy propellant oxygen in an external tank: indeed it requires none of the three tanks which launch the NASA Shuttle. It is hoped that the engine will be developed by Rolls Royce.

Whether such an ambitious project is even remotely possible depends, of course, on government policy and – money. It is certainly a revolutionary step in reusable spacecraft and if built would surely put Britain in the forefront of space technology. But the if, as always, is a huge one.

AN EYEWITNESS GUIDE TO THE SOLAR SYSTEM

Galileo explaining his conception of the solar system.

Opposite: *An astronaut at work outside the Space Shuttle, using the remote manipulator system.*

THE MOON
The King of Night

THE LANDING OF the Apollo 11 Lunar Module *Eagle* on the Moon on July 20 1969 and the subsequent exploration of its surface by astronauts Neil Armstrong and Edwin Aldrin enabled human beings for the first time to be more than just admiring gazers at the heavens – they became actual visitors to a new world.

That remarkable landing in the appropriately named Sea of Tranquillity is now a part of history, forever captured in the memories of those who watched from Earth and in photographs such as the ones on these pages. A new age of exploration had begun and Earth and Earthlings could no longer consider themselves isolated from the other bodies of the solar system.

Man had, in fact, dreamed of reaching the Moon – some quarter of a million miles away – ever since the days of antiquity when it was first worshipped as a god – 'The King of Night' according to the Slavs. The brilliant white satellite, some 2,160 miles in diameter and covered by huge plains and mountain ranges, was probably once closer to the Earth than it is now, but there is still a great deal of speculation as to just *how* it was originally formed. The alternatives are that it was a chunk of material that broke away from our own planet; that it was created from dust and gas left in orbit after the Earth's formation, or even that it was 'captured' while wandering through space.

Whatever its origins, one of its most striking features over the centuries

Opposite: *Earth-rise over the lunar horizon, photographed during the* Apollo 11 *landing.*

151

have been the eclipses caused when it passes directly in front of the Sun. The dark patches which dot its surface and are easily visible to the naked eye have also been an endless source of fascination. Once thought to be seas, but now precisely mapped and named geological features, they gave birth to the legendary idea of a 'Man in the Moon'. Also a constant source of delight are the changes it shows from the time of the 'Full Moon' to the slim and beautiful, 'Crescent Moon'.

Since that initial manned landing on the Moon by Armstrong and Aldrin 16 years ago we have learned a great deal about 'The King of Night', and a summary of the facts makes interesting reading after the generations of speculation and guesswork.

NASA's triumphant *Apollo* programme continued until December 1972, with further landings and extensive survey work, until it was finally halted for financial and political reasons. The Russians, though beaten in the so-called 'race' to the Moon, continued to send probes there for several years (including the *Luna* series which brought back a rich haul of samples), but they, too, eventually directed their space energies elsewhere.

Today, however, attention is once more being focused on our satellite – as much because NASA is making plans to go back there as because there are still a great many mysteries about it to be unravelled. And it appears that only by establishing a permanent lunar base will it be possible for scientists to solve these questions.

The *Apollo* missions brought back to Earth more than 2,000 samples of lunar rock, and NASA scientists have subsequently split these up into over 75,000 pieces from which they are still drawing dividends of information. For example, prior to the landings, scientists had established clearly enough that there was no appreciable atmosphere, but had not ruled out the possibility that there might be life forms of some kind. The samples dashed any such hopes by proving beyond any doubt that there had never been any water on the barren world. Only 'seas' of lava had ever existed – and then long ago in the Moon's relentlessly sterile 4.6 billion year history.

What the rocks *did* prove was that the Moon was rich in raw materials which could well be used by colonists from Earth. Unlike Earth rocks, they contain no water or organic materials but an abundance of metallic materials as well as oxygen. Research has shown they fall into two distinct groups.

Firstly, the light-coloured, rough-edged and much older rocks found on the highland areas which contain proportions of feldspar (a silicate of calcium and aluminium), pyroxene (a silicate of iron and magnesium) and in varying degrees amounts of potassium and phosphorus. Fragments of other rocks from meteors striking the Moon's surface over countless centuries have also been found among these samples.

The second group, from the plains, are much darker in colour, smoother, of feature and younger. They contain pyroxine, olivine, iron and titanium oxides, and are, in fact, a type of basalt which originated when the Moon's basins were flooded with molten lava from the interior.

However, both types of lunar rock contain at least three minerals formed only in the absence of water, and it is their abundance which scientists believe will prove so useful when ore-processing and air-generating plants are set up on the Moon. The silicon, too, has obvious uses – the micro-chip being the one which most readily springs to mind – while it is also believed the Moon soil can be processed into ceramics and glass, with the oxygen extracted from ilmenite to provide a vehicle propellant. Such resources, it is quite reasonably argued, would enable a Moon base to be established which could quickly become self-sufficient as well as providing an invaluable source of supply of minerals in the face of Earth's own declining resources. Because of the low gravity on the satellite (one sixth of that on Earth), it would also be possible to carry out the manufacturing processes quicker and more inexpensively than on our own planet. The Moon is, in effect, an almost perfect industrial environment.

The realisation of such facts has set NASA on course for the creation of a lunar base, and Hans Mark, a Deputy Director, believes a permanent colony of up to 2,000 people is quite feasible – and draws an interesting parallel with the way permanent bases were not set up in the Antarctic until some thirty years *after* Amundsen reached the South Pole in 1911. Here again the schedule for gaining a permanent foothold in a hostile environment is just the same.

Prior to this establishment of a base, however, it is believed a manned satellite will have to be placed in orbit around the Moon to first survey for suitable sites (seeking for the highest availability of raw materials) and then overseeing actual construction work. (It is hoped, too, that such an orbiting

observatory might also be able to provide answers to the strange puzzle of the 'hot spots' which have been detected on the Moon's surface. One theory which has been advanced is that they might just be caused by water existing as ice below the surface or in the darkness-shrouded craters in the polar regions.)

Because of the granite-like nature of the Moon's high landscapes it has been suggested these bases could even be hollowed out below ground, thereby providing the colonists ideal protection from the inclement

Above: *An artist's view of the solar system showing the orbits of the planets, the relative sizes of the planets and their approximate distances from the Sun, and the solar system seen looking towards Earth from the Moon.*

Top left: *Harrison H. Schmitt, one of the* Apollo 17 *crew, collecting rock samples in the Taurus-Littrow mountainous region of the Moon.*

Bottom left: *A beautiful full Moon, taken from* Apollo 17, *the last lunar manned flight, launched in December 1972.*

155

conditions of the surface. Such an operation would certainly be less expensive than a pre-fabricated lunar base, and with well-designed tunnels preventing any leakage of air or water, and sunlight ducted in by solar panels to provide light and power, the result would be a virtually indestructable and certainly long-lived subterranean colony.

On balance, however, most scientists seem to favour the idea of a lunar base along the lines which have been proposed by scientists like Wernher von Braun and writers such as Arthur C. Clarke – insulated, dome-shaped buildings accessible to the outside world through special air locks. From these buildings, lander spacecraft will be able to ferry the colonists to and from the orbiting satellites, while lunar vehicles will take parties on research and prospecting trips far across the surrounding landscape. It has also been strongly argued that the Moon, with its lack of atmosphere, will be an ideal spot on which to erect observatories to study the galaxy as well as a perfect jumping-off point for manned expeditions to Mars, Venus and even further afield.

As the idea of men and women living and working in controlled conditions on the Moon becomes ever more feasible, another scientist, Dr George Mueller, has looked still further ahead. For this former director of the *Apollo* programme believes that one day human beings may be able to live as naturally on the satellite as they do on Earth.

'Calculations have shown that a breathable atmosphere *could* be established on the Moon,' he explains, 'probably by chemical reactions on its surface for several hundred years, during which time its resources could be simultaneously developed.

'Later on the same thing might be done on other planets – though it would probably be more expeditious to move a moon out of orbit around somewhere like Jupiter and on to a collision course with Venus or Mars. This would quickly modify the environment of either planet. The excess of heavy gases would diminish on Venus, while the atmospheric gases and supplies of water on Mars would increase,' Dr Mueller adds.

From all these plans, it does not take a moment to realise that Neil Armstrong's first step on to the Moon was also the first step towards far greater achievements in space. And not just on that little world, but on the other planets of our solar system.

156

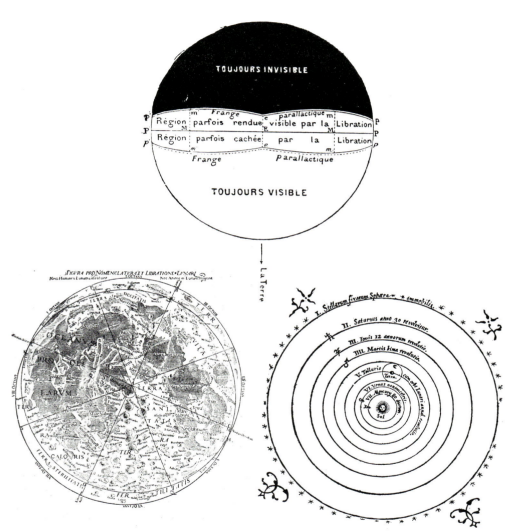

Top: *Early French conception of the Moon showing the visible side and the dark side.*

Left: *Mid-seventeenth century map of the Moon by the Italian astronomer, Riccioli.*

Right: *Facsimile of drawing of the solar system in a volume published by Copernicus in 1543.*

157

MERCURY
The Planet of Transits

As the world closest to the Sun – approximately 58,000,000 kilometres – it is no surprise to learn that Mercury has the harshest environment of any of the planets in the solar system. It is also, with Pluto, the smallest planet in the system; giving us the curious situation of the innermost and outermost of our worlds being identical – though very much the opposite in temperatures! And because one side of Mercury is always facing the Sun and the other side away from it, it was for years thought wrongly to be both the hottest and coldest place in the universe.

The brightness of our sky prevents us from seeing the planet with the naked eye except just before sunrise and just after sunset when it appears momentarily near the horizon. For this reason only a handful of people have been able to claim to have seen the planet, and pictures of it have remained very rare.

This situation was to a degree remedied when the NASA spacecraft *Mariner 10* flew by Mercury for the first time in March 1974. It gave scientists and astronomers the first real opportunity in history to look at a world which is so difficult to study by conventional techniques because it is never more than 23 degrees from the sun's blinding light.

The fly-by, within a few hundred kilometres, provided some striking images of what proved a very sombre world and revealed that it had a surface not unlike that of our Moon and an interior which very strongly resembled

Opposite: *A simulated Mercury encounter by the NASA* Mariner *Venus/Mercury spacecraft.*

our own! The pictures showed a brown-hued surface pock-marked with craters, while the craft's measuring instruments indicated there were large temperature variations from a high of 430°C at noon on the equator to about 170°C at midnight in the same area.

Mercury was confirmed to have a diameter of 4,878 kilometres and a rotation period of 59 days. With an orbital period of 88 days travelling around the Sun, it can be seen to rotate just three times while circling the Sun twice. *Mariner 10* also informed the NASA scientists that the planet had no discernible atmosphere, with a force of gravity twice that of the Moon. Of life forms there were unquestionably none.

Such information put to an end many of the mysteries about Mercury which had persisted since the very earliest times. Despite its obscurity from human eyes, it was known as a planet in Sumerian times in the third millenium B.C. In Classical Greece, however, it was believed to be *two* stars, one a morning star called Apollo, and the other an evening star called Hermes. It was this Greek equivalent of the Roman god, Mercury, which led to it being given the name by which it is familiar.

Nicolaus Copernicus, who devised the theory of planetary motions before

Above: *Seventeenth Century painting symbolising the transits of Mercury.*

Opposite: *Photo mosaic of Mercury as seen from* Mariner 10 *during its approach to the planet. The inset shows a bright crater on the edge of a larger crater. The bright crater was named in honour of the late Dr Gerard Kuiper, a member of the* Mariner 10 *team.*

the invention of the telescope, was greatly intrigued by the planet, and is said to have declared on his death bed that the only thing he regretted in his life was not having seen Mercury. The most famous pictures of the surface features (before *Mariner 10*, that is) were probably those of the German astronomer, Johann Hieronymous Schroter, made in the early nineteenth century.

The 'Transits of Mercury' are, of course, famous – and are caused by the planet crossing the face of the Sun about fifteen times per century. Viewed from Earth, this phenomena which has been recorded since the year 1677, rather resembles that of a Sunspot. As a matter of interest, the next one is due on November 12, 1986.

Although, as I mentioned earlier, the planet possesses a great many large, well defined craters on both hemispheres mostly three to four billion years old, those on the left side are the most pronounced. Of these, the 40 kilometre crater named Kuiper is clearly visible when reflected in the Sun's rays. The biggest is a huge basin, Caloris, nearly 1,400 kilometres across, which *Mariner 10's* penetrating camera eye established was almost certainly caused by a large impact.

The mission also confirmed that Mercury has a large iron core, and that the

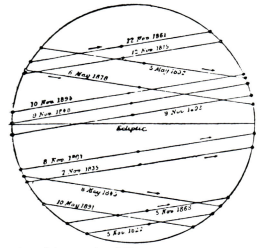

Nineteenth century drawing of the transits of Mercury.

cooling and contraction of this during the planet's infancy was undoubtedly the cause of the other most prominent features on the surface: gently undulating cliffs, some of them running for many hundreds of kilometres.

But despite this iron core, scientists studying the spacecraft's reports were forced to the conclusion that the planet has a weak magnetic field – perhaps as little as one per cent of the strength of the Earth's. They were puzzled, too, by the fact that this core is aligned *with* the axis of the spin, making them wonder just how it could rotate at all! This strange anomaly is encountered nowhere else in the solar system.

It would seem further missions by orbiting spacecraft are essential to find an answer to this mystery – as well as giving us still further information about the strange almost skeletal appearance of this innermost of our worlds.

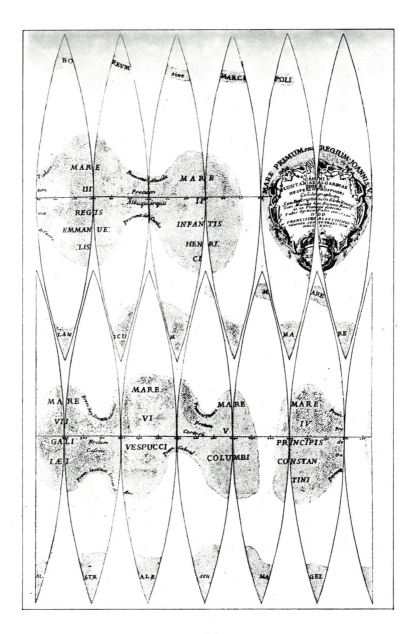

164

VENUS
The Sister of Earth

VENUS WAS the first planet in the solar system onto which a space vehicle from Earth was landed. This occurred on March 1, 1966 when the Russian space vehicle, *Venera 3* touched down on the surface of the world that has often been thought of as 'Earth's sister planet'. After the Sun and Moon, Venus is the brightest natural object in our skies as well as being the world which comes closest to us – a matter of 26,000,000 miles.

The first fly-by of the planet had been made three years before the Russian landing by the US *Mariner 2* in December 1962, but the Russians were certainly the most keen observers of Venus until the *Pioneer* orbiters of the late 1970s. Yet despite all the study by both groups there are still a great many mysteries remaining to be solved.

The mystery which has perhaps puzzled observers more than any other over the centuries has been the dense and impenetrable blanket of yellowish cloud which swathes the planet to a height of at least 100 kilometres above the surface. Thanks to the probes, however, it has been possible to establish that this atmosphere is almost 95 per cent composed of carbon dioxide, with small traces of sulphur dioxide which probably gives the clouds their yellowish colour as distinct from Earth's white clouds consisting of water and ice. There are also traces of water vapour.

Perhaps even more importantly it was discovered that Venus – like Earth – has in its atmosphere large concentrations of Argon, the gas which was a vital

Opposite: *Italian map of Venus, dated 1727.*

Artist's conception of Ishtar Terra, the highest and most dramatic highland region on Venus, based on topography measurements by the Pioneer Venus Orbiter *spacecraft.*

element of the primordial solar nebula from which the planets and their atmospheres were formed.

The surface of Venus is, however, inordinately hot – as much as 470°C – almost three times the highest temperature ever measured on Earth. Curiously, the probes have shown that this does not vary greatly anywhere on

166

the surface of the planet, although it does appear to be a little warmer above the poles than on the equator!

Venus also turns very slowly – rotating once in 243 days and in an *anti-clockwise* direction: the reverse direction to all the other planets going round the Sun! No-one is quite sure why it should behave in this way, though it has been suggested that the Earth's gravitational pull may be responsible. The planet takes 225 days to orbit the Sun, 108,200,000 kilometres away. A Venusian day lasts for 2,800 hours or the equivalent of 120 Earth days.

The 12,100 kilometre diameter world has been the subject of intense interest for mankind since the very earliest times. As early as the year 3100 B.C. it was associated with the Goddess of Love, although in some pieces of classical literature it is identified with Lucifer and its glowing surface is said to personify Hell!

Simulated colour representation of the highlands of Venus. Lower elevations are shown in blue, medium in green, and high elevations in yellow.

The earliest scientific speculation on the environment of Venus appears to have been carried out by the Dutch scientist, Christiaan Huygens, in the seventeenth century. And it was a Russian chemist named Mikhail Lomonosov who suggested in 1976 that the planet had an atmosphere at least as dense as that of Earth.

It has taken until recent years and the cameras of the space probes to finally penetrate Venus' dense atmosphere and reveal beneath it a dramatic and awesome landscape still evidently in the process of formation. Almost sixty per cent of the planet's topography consists of flat, rolling plains, scattered with rocks and a fine dust which the *Pioneer* 'day probe' briefly disturbed on landing. (Despite this spacecraft supposedly only having a resistance to temperatures up to 200°F, it in fact survived for some 67 minutes on the

The Pioneer *Venus Probe which landed on the surface of Venus in 1978.*

burning wasteland until finally ceasing transmission when the temperature inside the titanium vessel rose to 260°F!)

The orbiter also sent back data on a number of craters on the surface, as well as a gigantic plateau almost as large as the continent of America. It pinpointed four major highland regions, too, the highest of these being dominated by a massif bigger than Mount Everest and towering to almost 39,000 feet. It has been named Ishtar Terra after the Assyrian goddess of love and war. To NASA scientists these gigantic ranges seemed somehow most appropriate on a world with an atmosphere 90 times more massive than Earth's.

Evidence provided by the *Pioneer 12* probe seems to confirm the belief that Venus has some of the largest and most active volcanoes in the solar system, and two probable sites of active cones have been mapped at sites named Atla and Beta Regio about 4,800 miles apart.

Despite this activity, erosion obviously takes place very slowly on the planet, as images taken by two of the Russian space probes, *Venera 9* and *10* which landed on the surface in October 1975 clearly showed. Three years later, in December 1978, *Venera 12* picked up crackling sounds in the atmosphere and this has led scientists to the view that lightning occurs at lower altitudes on Venus, and that the world may well be frequently illuminated by lightning flashes. There is strong evidence, too, of thunderstorms and powerful winds disturbing the dense clouds. A further curious feature brought to light by a recent *Pioneer* mission has been evidence of what may well be 'holes' in the atmosphere just above the north and south poles.

Facts such as these have lead to the feeling that anyone standing on the surface of Venus would probably notice scattered patches of sunlight rather like an overcast day on Earth. The refraction in so dense an atmosphere would also be so strong that an observer would be able to see beyond the horizon as the surface level appears concave. Indeed, the viewer would always seem to be in a depression wherever he stood!

What does, though, continue to puzzle all scientists is why two neighbouring planets like Venus and the Earth should have atmospheres that are so different when they must have been so similar in their formative states. As Venus is closer to the sun than us, it has been suggested that the heavy cloud formation has created a 'greenhouse' effect with the sun's rays burning up the surface. And this naturally prompts the question: could evolution bring about the same fate for us if we allow the protective screen provided by our atmosphere to be weakened through careless misuse of the environment which sustains it?

Intriguingly, it has also been put forward that man might consider reversing the process on Venus by the introduction of green plants that would turn the carbon dioxide into oxygen. The major problems, of course, are the heat and the lack of water to generate growth – but no one is yet prepared to completely write off the idea we might one day find ways of establishing a colony on our sister planet.

In the meantime, we can await with interest the results of the latest space mission to Venus being conducted by the Soviet Union. Although few details have been released, on December 15, 1984 they launched *Vega-1* space probe which is scheduled to deliver a landing module to the planet some time in 1986. Once again, if it survives it will help build up still more our knowledge of the solar system in which we live.

Above: *A nineteenth century impression of Venus.*

Opposite: *Venus taken by* Mariner 10's *television cameras in invisible ultraviolet light.*

MARS
The Red Planet

WHEN THE American spacecraft, *Viking 1* settled down safely on the Chryse Planitia landing site in the northern hemisphere of Mars on July 20, 1976, after its year-long half a billion mile journey from the Earth, a scientist at the Jet Propulsion Laboratory in California who had been monitoring the flight was heard to say as the first pictures began to be transmitted from the rock-strewn surface, 'I almost expected to see camels!'

The date of the landing was exactly seven years to the day after *Eagle* had touched down on the Moon, and equally it was to open up a whole plethora of new information on a world that held a similar place in the interest of mankind. The Red Planet – as it has been popularly known for centuries – inspired this interest as much as anything because it is the only planet in our solar system that bears any real resemblance to our own home. Even early astronomers could distinguish white capped poles, dark, blue-grey formations in constant stages of change, and the appearance of valleys, mountains and rivers. Small wonder that it seemed to man that here might be another inhabited world like Earth – and so he gave Mars a special place in his imagination and in his literature.

The *Viking* landing on Mars actually climaxed a long and far from trouble-free series of attempts to establish a beach-head on this intriguing world. Although the first space probe to fly by Mars had been *Mariner 4* in July 1965, it was not until November 1971 that attempts were made to land on the surface. The primary reason for this was, of course, that setting down in the Martian atmosphere was far trickier than landing on the airless Moon.

The Russians, in fact, have tried to land on Mars four times, twice in 1971 and twice again in 1974. In the first instance in 1971, one lander crashed and the other stopped sending data back after only 20 seconds of transmitting. Three years later, the first of the pair of landers inexplicably flew straight past the planet, while the instruments on the second again failed – this time on the descent to the surface. To their credit, they have not been deterred and now plan to send a probe to one of the Martian moons, Phobos, in 1988 – perhaps as a preliminary to a manned voyage to the planet itself.

NASA was more fortunate with *Viking 1* in 1976, although there were moments of high drama as the lander appeared in danger of toppling over when it put one leg on a rock. Two months later it was joined on the surface by *Viking 2* which made a landfall on the opposite side of Mars amidst the dunes and volcanic rocks at the site named Utopia Planita. Between them, these two machines then began to satisfy mankind's curiosity about a world known since Babylonian times over 3,000 years ago.

Because of its bright ochre red appearance, Mars was associated with destruction in the mind of the Babylonians who called it Nergal after their god of death and pestilence. The Greeks named it Ares, for their god of battle, but it was the Romans who settled the issue with Mars – God of War. Predictably, the planet's two moons shared the same association being named after the mythical chariot horses of Mars – Phobos (Fear) and Deimos (Terror).

The planet is 227,900,000 kilometres from the Sun (just over one hundred and forty million miles) and orbits it once every 687 days. The radius around the equator is 6,787 kilometres (about half that of Earth), but it enjoys a day almost exactly the same as ours (24.6 hours).

Because of its seemingly erratic motion, Mars was something of an enigma to the early astronomers, and it was not until the start of the seventeenth century that Johannes Kepler laid the foundation of serious study by discovering that the planet's orbit was an ellipse rather than a circular one along which it moved with non-uniform but predictable motion. Since these conclusions were arrived at, Mars has played a major role in the development of planetary astronomy.

Galileo made the first telescopic observations in 1610, noting that Mars was not perfectly round, but it was a quarter of a century more before the first

Above: *This computer enhanced Martian sunset was taken in August 1976 by* Viking 1. *The blue to red colour variation is caused by a combination of scattering and absorption of sunlight by atmospheric particles.*

Top left: *Colour-reconstruction photograph of Mars taken by the* Viking 1 *Orbiter in June 1976.*

Bottom left: *This view of Mars shows the smooth plain, called Argyre, a large impact basin which has been observed from Earth for many years.*

drawings to show markings on the surface were made by the Italian astronomer, Francesco Fontana. In 1659 the redoubtable Dutchman, Christiaan Huygens, produced the first accurate drawing, and this was followed by increasingly more complex versions by the German astronomers, Wilhelm Beer and Johann H. von Mädler (1830), the French scientist Camille Flammarion (1877), and two Englishmen, Richard A. Proctor in 1867 and Nathaniel E. Green in 1877.

It was the American astronomer Percival Lowell who identified what he thought were 500 'canals', insisted that they must be the work of intelligent beings, and began the speculation which has dominated so much of the discussion about Mars being inhabited. It was another American, Asaph Hall of the U.S. Naval Observatory, incidentally, who discovered the two moons, Phobos and Deimos in 1877.

An immediate discovery that *Viking 1* could confirm on landing was that the atmosphere of Mars is much thinner than Earth's – a hundred times less dense in fact – and 95 per cent carbon dioxide with about three per cent nitrogen. (By way of comparison, let me say that a human would need to climb 21 miles up in our atmosphere to experience the atmospheric pressure that exists on the surface of Mars – and in such thin air his blood would boil!) It also transpired that the planet is too cold (average temperature −50°C) and the atmospheric pressure too low for water to exist for long as a liquid – instead it changes directly from vapour to ice and back again.

However, this does not automatically preclude there *ever* having been water there, for based on the *Viking* data, NASA scientists are of the opinion that a billion or more years ago heat from within the planet melted a huge area of sub-surface ice (like the permafrost in the Earth's arctic regions) and caused cataclysmic floods which carved channels and gorges across the surface, as well as tablelands, craters, cliffs, pits and even the tear-drop shaped islands in what look like old stream beds. All the terrain, they believe, bears the hall-marks of having been sculptured by the force of water.

It was, though, the photographs which *Viking* transmitted to Earth which caused the greatest excitement. They were – and are – masterworks of unearthly beauty: many of them disclosing for the first time colossal formations on the surface of the planet.

For instance, not far from the Martian equator was revealed a huge

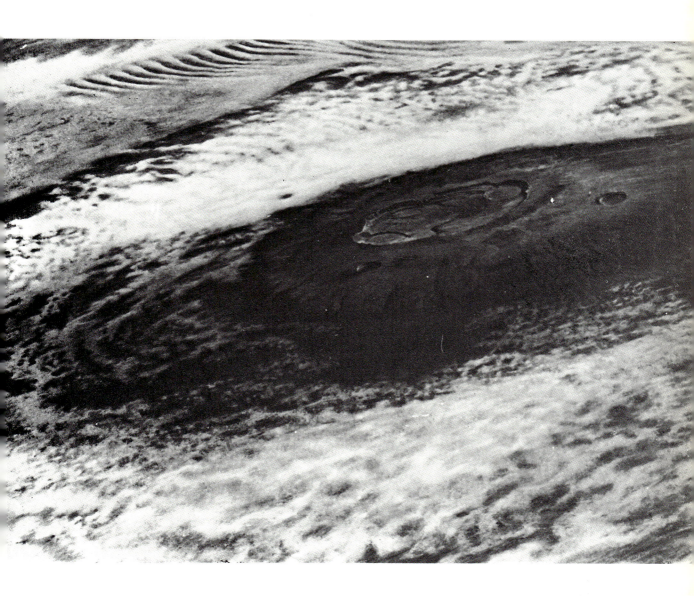

The huge volcano, Olympus Mons, is fifteen miles high with a base of almost 335 miles.

volcano, named Olympus Mons, which could be seen at a glance to be three times as high as Mount Everest. Measurement indeed proved it to be fifteen miles high, with a base of almost 335 miles, and a multi-ringed crater all of 45 miles from one side to the other. Equally breath-taking was a huge canyon system, Valles Marineria, situated just below the equator line. Four miles deep, 150 miles wide, the system straggled across an area the size of the United States!

To our eyes, the pictures from *Viking* seem to paint a landscape no less

Deimos, smaller of the two satellites of Mars.

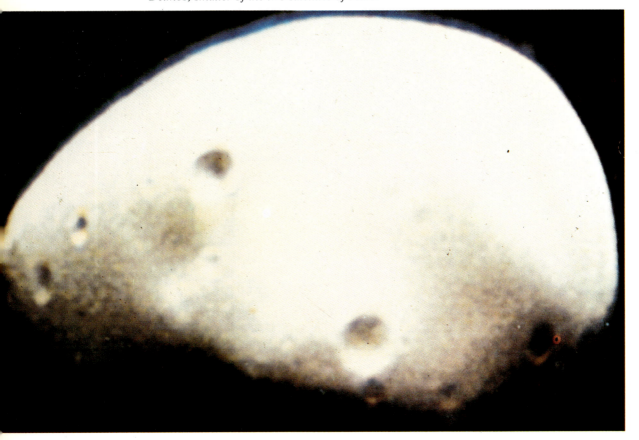

The heavily-cratered surface of Phobos, larger of the two Martian satellites.

hostile than some of the Earth's great deserts; red, dusty and rock-strewn. It was only the other data from the scientific instruments which pointed out just how the Sun's ultraviolet rays were searing the landscape in a manner quite different to our own environment where we are shielded by the ozone in our stratosphere.

With the passing months, the two landers began to relay a stream of information about the changing seasons of the Martian year which, as I mentioned earlier, lasts for 687 days. The weather, it was shown, varies greatly not only with the season, but also the time of day. On a 'summer' day,

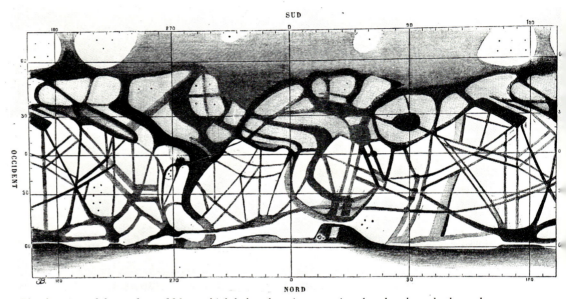

The drawing of the surface of Mars which led to the misconception that the planet had canals.

for example, the temperature can go from −123°F at dawn to a 'high' of −24°F at noon! In the winter, massive temperature differences between the equator and the ice-capped poles produce brisk westerly winds and intense low pressure areas not unlike our own terrestrial systems.

Viking also provided data on the extraordinary Martian winds. These are generated by the Sun's rays heating particles of dust in the air, and the storms which result can reach speeds of up to 250 miles per hour! In 1977, for example, the landers recorded 36 of these dramatic storms – two of which were so vast they turned into global disturbances!

The Martian clouds were found to fall into four main categories: convective clouds, wave clouds, orographic clouds and fogs. During its observation, *Viking 1* provided some of the most amazing weather pictures ever seen when it transmitted shots of some early morning fog developing in low-lying areas as the frost was vapourised by the early morning sun. In the evening, it showed the fog returning directly to ice once more.

180

The NASA missions have also provided new details on the planet's moons, Phobos and Deimos. Phobos, the inner and larger of the two, lies very close to Mars (9,380 kilometres away), and has a unique orbit of seven hours thirty-nine minutes which means that it travels around the planet *twice* in a Martian day!

They are both ellipsoidal satellites and have dimensions of 27 kilometres and 12 kilometres respectively. However, they are noticeably different in that the surface of Phobos has numerous grooves as much as 200 metres wide and a large crater named Stickney about 10 kilometres in diameter, while Deimos is pock-marked and littered with loose fragments of material. Despite their differences, scientists believe both moons are made of meteoric material and were probably originally asteroid fragments captured by Mars' gravity field as they hurtled by.

And finally to the question most scientists and laymen want answered about Mars: does *any* form of life exist there? Even to find microbes of some sort would mean that we were not alone in the universe. Dr Michael McElroy who has studied the *Viking* data puts the facts this way:

'The elements in the chemistry set are there,' he says. 'We have water, we have carbon, we have nitrogen, we have sunlight. The only real remaining question is whether the Great Chemist was there putting the elements together in the right way.'

No-one believes life forms of the kind imagined in Science Fiction novels such as those of H. G. Wells, Edgar Rice Burroughs and Ray Bradbury will be found on Mars, but some scientists think that microbes *are* a possibility. One of these men is Dr Norman Horowitz, a Californian biologist.

'From a chemical point of view, there is only one form of life on Earth,' he explains. 'We will next want to know whether the Martian organism, if it exists, is made the same way we are – with the same DNA, the same genetic code, proteins made of the same amino acids. By answering those questions we can learn a great deal about the origins of life and its abundance in the universe.'

With such a possibility held up before us, is it any wonder that both the American and Russian space agencies have visits to Mars – both manned and unmanned – firmly focused in their future schedules? The 'race' to Mars – just like the earlier dash to the Moon – could be on very soon.

The first evidence of a ring around Jupiter was seen in a photograph taken by Voyager 1 *in March 1979. In this photograph a line has been drawn around the planet to show the position of the extremely faint ring.*

JUPITER
The Giant Planet

JUPITER, THE most massive of all the planets in the solar system, was named by ancient astronomers after the ruler of the gods in the Graeco-Roman pantheon. They had no real idea of the planet's dimensions and regarded it with a sense of awe which was, if anything, heightened when the spacecraft, *Voyager 1*, brought it into close focus for the first time in December 1979 after a 422 million mile journey. (There had, in fact, been earlier fly-bys by *Pioneers 10* and *11*, but with only limited results.)

This enormous world – over 1,300 Earths would fit into its bulk – complete with its famous Red Spot, dazzled viewers when *Voyager* began transmitting pictures back to Earth – but also amazed them with two heretofore unknown facts. Firstly, that Jupiter has a circumnavigational ring of fine rock fragments 36,000 miles above the cloud tops. And, secondly, while studying the little moon, Io, an umbrella-shaped plume was seen to rise into the air, indicating to the scientists back on Earth the remarkable news that for the first time another world had been discovered in our solar system that was geologically alive.

Jupiter is, in fact, the first of a group of four planets beyond the asteroid belt which are markedly different from what we call the terrestrial planets of Mercury, Venus, Earth and Mars. This group are called the Jovian planets and consist of Jupiter, Saturn, Uranus and Neptune. They are huge, rapidly-rotating, low-density worlds which despite dense atmospheres are composed of light elements just like stars. Indeed, Jupiter emits nearly twice the energy it receives from the Sun and so is rather more like a failed star with its own internal heat source that a typical planet. This additional energy source not surprisingly has an important role in the planet's meteorology as its weather is affected both by the internal heat source and the external solar radiation. Consequently the weather systems on Jupiter are unique – quite unlike those of any other planet.

The planet, with its alternating light and dark belts of atmosphere running parallel to the equator, is totally shrouded in swirling clouds, and unlike the Earth with its single zone of weather based on the evaporation, condensation and precipitation of water, Jupiter has *three*. One zone of water, one of ammonium hydrosulfide and one of ammonia. There are also small quantities of other strange components such as methane, acetylene, phosphine and germanium terahydride, in and around the cloud tops, and it is now believed that in a complex process of atmospheric chemistry they play a part in the creation of the colourful and exotic appearance of the planet which one NASA scientist has described as being 'like bubbling bright paints that will not mix'.

Below the atmosphere, the planet is mainly liquid and helium with no solid surface as such. This discovery confirmed for scientists a theory that Jupiter was originaly most likely formed from a swirling mass of dust and gas and can, in fact, be compared to a gigantic 'bag' of gases consisting of hydrogen and helium in vast 'shells' tens of thousands of kilometres thick around a molten core of silicates and metals. (Originally, say the scientists, the flaming red ball may have been as much as 200,000 kilometres wide, but has since shrunk to 143,000 kilometres.) The core of Jupiter is probably just about the same size as our Earth, and though the planet has 318 times the mass of Earth, its density is only one quarter! In essence, then, the composition of this giant world is much like that of the Sun.

Jupiter rotates faster than any other planet in our system – once in just under ten hours – but because of its distance from the Sun (788 million kilometres) its year is equivalent to just under 12 Earth years. There are enormous temperature variations, too, ranging from −130°C at the cloud tops to anything up to 54,000°F near the core! The thermal radiations from the planet have helped create the huge magnetic field around it in which float its covey of sixteen moons: the four major Galilean satellites, Io, Europa, Ganymede and Callisto, to which we shall return in a moment, and the other twelve much tinier moons. Moving outwards from Jupiter these are: Un-named (designated 1979 J3), Adrastea, Amalthea, Un-named (designated 1979 J2), Leda, Himalia, Lysithea, Elara, Ananke, Carme, Pasiphae and Sinope.

During its fly-by, *Voyager* relayed the information back to Earth that all the moons are being bombarded by lethal electrons which means their surfaces are constantly being changed and eroded by this enormous power. The spacecraft also recorded that lightning crackled continuously in the Jovian atmosphere: and this has led to speculation that the energy could be triggering the formation of organic molecules – the chemical foundations of life. And further that the orange luminous meteoric phenomena of electrical character which we know as the Aurora, is far more intense on Jupiter than had been previously expected. In fact, on the dark side of the planet an auroral arc some 30,000 kilometres in length was observed – the largest yet seen on any body in the solar system.

Voyager also sent back a series of high definition photographs of the remarkable cloud formations as well as some stunning pictures of the Great Red Spot. The Jovian clouds could be seen to be of varying sizes, colours and shapes, and it was observed that though they seemed to be in a constant state of turbulence, they lasted for considerably longer periods than similar kinds of formations on Earth, probably brought about by an unusually slow dissipation rate.

The pictures of the Great Red Spot in the southern hemisphere were breathtaking. Measuring 21,000 kilometres long by 11,000 kilometres high, this unique spot which is big enough to hold two worlds the size of Earth, has fascinated astronomers ever since it was first noticed in 1664 – but it took *Voyager* to establish that it is, in fact, a high pressure point in the atmosphere – one of several such vortices on Jupiter – and its vivid colour is probably caused by an excess of phosphine.

While *Voyager's* eyewitness report on the planet answered a number of long-standing questions, it also left just as many unanswered, in particular whether life – or its chemical precursors – might be found there. If the answers are to be found they may well lie in further study of the atmosphere, for scientists are beginning to wonder if the many different colours of the planet might just be the work of organic molecules.

Not far from the 'surface' of Jupiter lie the four larger moons which have been named the Galilean satellites after Galileo who discovered them in 1610. Curiously, although Galileo named them after legendary figures closely associated with Jupiter (Zeus) in Greek mythology, he was actually going to call them collectively the Medicean stars in honour of his patron, Cosimo de Medici. It was the astronomer's contemporary and great rival, the German Simon Marius, who insisted on them being named after their discoverer.

The four moons were actually the first bodies in the solar system to be

Below: *This photograph by* Pioneer 10 *clearly shows the Great Red Spot and never-before-seen details of the giant planet's cloud tops.*

Opposite: *Jupiter and its four planet-size moons, the Galilean Satellites. This artificially composed collage shows the satellites in their relative positions but not to scale. Reddish Io (upper left) is nearest Jupiter, then Europa (centre), Ganymede and Callisto (lower right).*

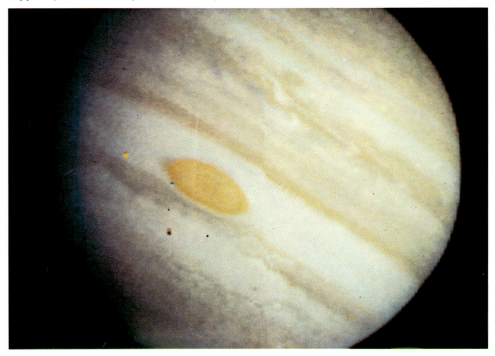

discovered by telescope, and they somewhat resemble a mini solar system themselves. Indeed, a number of scientists believe that if we can discover how *their* creation came about, this may well solve the problem as to how our own solar system was formed. *Voyager* helped take a giant step along this path with its dramatic photographs, for prior to the spacecraft's fly-by we knew little more about Io, Europa, Ganymede and Callisto than their appearances, sizes and densities. Then, suddenly, there they were in full colour close up.

Io is the innermost of the group, orbiting about 421,600 kilometres from the planet and named after the unfortunate girl of mythology who, after her romance with Jupiter was turned into a heifer. The moon is some 3,640 kilometres in diameter (much the same as our own Moon, in fact) and most probably consists of a solid core with a molten silicate interor and sulphur and sulphur dioxide crust – giving it the appearance of a vivid red desert world.

A surprising piece of information that *Voyager's* instruments picked up was that a powerful electric current was evidently flowing between Jupiter and Io. This power was estimated at two trillion Watts – or roughly the combined capacity of every single power plant on Earth! Study of this phenomenon led the NASA scientists to believe that it was directly responsible for the hyperactive state of the moon and the volcanic eruptions it was giving off, some of them hundreds of miles high.

The very first time one of these sulphur eruptions was seen was on March 9, 1975, when an observer at the Jet Propulsion Laboratory in California suddenly caught sight of a plume rising to 280 kilometres in the atmosphere. It was a moment of history – the first evidence of active volcanism beyond the Earth and a clear indication that another body in our solar system was geologically alive. In the days which followed, *Voyager's* cameras spotted eight more active volcanoes which confirmed the first impression that although the satellite might be totally arid on the outside, it throbbed with volcanic activity internally and was probably the most intensely heated terrestrial body in the solar system!

Further study of the data supplied by the spacecraft has shown that Io is seemingly caught in a tug of war between the mother planet and two of its sister moons, Europa and Ganymede, which literally cause its surface to rise and fall by a hundred metres and more every day!

Although Io evidently has no wind, the volcanic eruptions give off constant 'falls' of yellow, orange and blueish-white dust which covers everything and also creates a foul-smelling 'atmosphere' not unlike that to be found in the volcanic regions of Earth. A landing on Io would, in fact, be rather like a journey to the heart of the most violent eruptions of Mount Etna.

Europa, which lies next in orbit, is quite different to Io but equally full of surprises. Just over 3,000 kilometres in diameter, it circles Jupiter at an average distance of 670,900 kilometres, and has a silicate interior beneath a crust of ice. The pictures *Voyager* transmitted gave the overwhelming impression that here was a world sunk deep in an ice age, and in all probability the pack-ice surface is only broken by the occasional area of 'ponding' where water has broken through the surface along with the occasional geyser showering ice pellets into the air which would be far more unpleasant to experience than any sleet storm on Earth.

Without doubt, though, Europa is the smoothest body that has ever been seen in the solar system, showing no evidence whatsoever of any topographical features. Only a vast network of cracks cover the surface of this moon appropriately named after the gentle maiden who was seduced by Jupiter and then carried off to Crete to become an object of worship. And like that enigmatic beauty from mythology, unique Europa poses more questions about its nature at the moment than Earth scientists can answer. One puzzled NASA man has accurately described it as 'looking like a cracked white billiard ball – and just as devoid of ready answers.'

Ganymede, named after the handsome youth that Jupiter sequestered to become cupbearer to the gods, is the largest of the four Galilean moons at 5,270 kilometres in diameter. It circles the mother planet at just over one million kilometres and has an icy crust over a silicate core. As huge in size as the planet Mercury, its surface ice is much deeper than that on Europa, and NASA scientists believe the moon may once have had huge separate areas of ice rather like the continental plates on Earth. These Tectonic Blocks, as they are called, may have constituted land masses as we understand the term. Certainly the almost dirty-brown features of Ganymede reveal on closer inspection to have craters, ridges and mountainous regions, while the alternating parallel grooves which can also be seen indicate crustal movements billions of years ago.

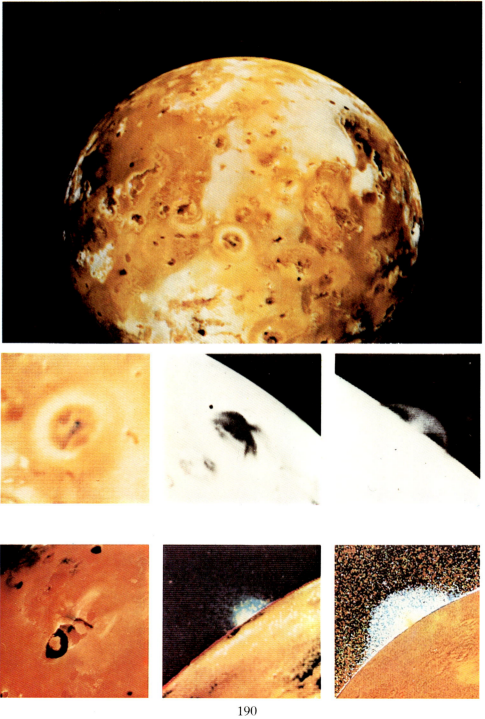

Top left: *This view of Io was taken on 4 March 1979, from a distance of 500,000 miles, by* Voyager 1.

Bottom left: *Views of the active volcanic plumes on Io from* Voyager 1.

Above: *Colour image of Europa, smallest of Jupiter's four Galilean Satellites. Europa has a density slightly less than Io, suggesting the presence of a substantial quantity of water.*

Top: Voyager 2 *colour photo of Ganymede, the largest Galilean Satellite. The bright spots dotting the surface are relatively recent impact craters.*

Another interesting feature about Ganymede which *Voyager's* all-seeing eyes spotted were splashes of vivid white at various points on the surface. These are in fact patches of new white ice thrown up at the impact of meteors striking the little satellite. A visit to this frigid world would be incomparably more demanding than a trek across the Earth's Arctic Circle.

Callisto, the last of the four Galilean moons, orbits Jupiter at an average distance of 1,880,000 kilometres. It is 4,850 kilometres in diameter and once again has a silicate core covered by a soft ice mantle and a thick ice crust. Named after the beautiful young girl who lured Jupiter away from his wife, Juno, and was later turned into a bear by the enraged spouse, the satellite's most notable feature is its deeply cratered surface. Indeed, it has the most densely cratered surface in the entire solar system – although *Voyager* enabled us to see for the first time that these craters are actually quite shallow and lack any sharp features. This would seem to indicate that they are extremely old.

The spacecraft also spotted a giant, 600-kilometre-wide crater at the heart of numerous concentric rings stretching for at least 1,000 kilometres in all directions. This, the NASA scientists agreed after many hours poring over the photographs, must surely have been caused by some cataclysmic-sized meteor striking the moon at an earlier date in its history. Callisto must, in fact, have received constant bombardment by meteors throughout its four and a half billion years of existence, and but for the ice slurry that has partially obliterated the craters, would appear an even darker and more moribund place than it is.

All this said, Callisto remains the only world in the Jovian system on which mankind might feasibly land at some time in the future, NASA observers believe.

As it passed Jupiter and its satellites, *Voyager* also gave man his first eyewitness view of the tiny moon, Amalthea, circling close to the planet once every twelve hours and regarded as little more than a tiny point source since its discovery less than a hundred years ago in 1892. The cameras revealed that it is, in fact, a fragment of material about 130 kilometres long by 170 kilometres high, and oblate in shape rather than spherical. This factual explanation laid to rest some rather imaginative theories that it might be an asteroid or even an abandoned space machine!

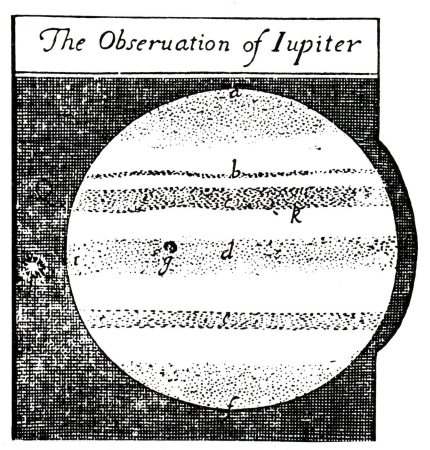

Jupiter, drawn by Robert Hooke, as it appeared on 26 June 1666.

The extremely dark red colour of Amalthea (another mythological name) has been seen by this closer examination to have numerous craters on its surface – and indeed the strange shape of the satellite is now generally believed to have been caused by innumerable impacts.

Such has been the excitement caused by *Voyager's* 'report' from Jupiter, that NASA is now advancing plans for an even more ambitious unmanned space probe to fly there in 1986. Appropriately named *Galileo*, this craft will be launched from a Space Shuttle, and will consist of two units which will separate as they near the giant world. One part, the larger, will place itself in orbit to observe and photograph the planet and its moons, while the smaller section will plunge directly into the atmosphere. Although it will be protected by a carbon heat shield, the probe is only expected to survive for about an hour in the intense heat – but in that precious hour it will relay back a multiplicity of fresh information. In particular, whether some of the warmer regions of the planet, which are believed to resemble Earth in its prehistoric age, might *just* support life.

Although most scientists believe the atmosphere of Jupiter is a little too unstable for life, there are those who are prepared to speculate on the idea of life forms existing in the swirling clouds looking something like giant jellyfish. Whether *Galileo* can confirm such a fascinating idea or not, it will certainly enable mankind to take yet another step closer towards understanding the nature of awe-inspiring Jupiter and its covey of intriguing moons.

Opposite: *This photograph of Callisto, taken by* Voyager 2 *on 7 July 1979, shows its heavily-cratered surface. The bright areas in this ultraviolet-imaged photo are recent impact craters.*

SATURN
The Gem of the Heavens

AFTER THE TWO *Voyager* spacecraft had completed their fly-bys of Jupiter, NASA scientist Thomas A. Mutch remarked, 'We had known that Jupiter would be remarkable, for man had been studying it for centuries, but we were far from prepared for the torrent of new information that the *Voyagers* poured back to Earth . . . it was like being in the crow's nest of a ship during landfall and passage through an archipelago of strange islands.' What the craft found when they passed the next planet, Saturn, was to prove, if anything, even more remarkable.

This huge, ringed world hanging in the heavens like a glowing yellow ball is arguably the most easily identifiable and romantic-looking planet in the entire solar system. Like Jupiter, Saturn had come under scrutiny during the mission of *Pioneer 11* in September 1979, but it was *Voyager* that brought home the true impact and beauty of this world in November 1980.

For centuries, Saturn was believed by the ancients to be the outermost of all the planets, and Galileo, the first man to observe it by telescope in July 1610, actually thought it was not one world but *three*! The great astronomer did not recognise the ring system for what it was and thought he had found a triple planet – three worlds almost touching one another. It was the Dutchman, Christiaan Huygens, who accurately described the phenomenon

Opposite: *The remarkable, huge ringed world, Saturn, photographed by* Voyager 2.

197

in 1659. 'Saturn is surrounded by a thin, flat ring,' he said, 'which nowhere touches the body.'

This lovely planet, some 947,000,000 miles from Earth, was certainly known to the Babylonians as early as the mid-seventh century, but it was the Greeks who first named it after Cronos, the original ruler of Olympus, whose Roman equivalent is Saturn, the god of Time. Some 120,600 kilometres in diameter and rather flatter at the poles than the other planets of our system, it revolves swiftly (a day takes just over 10 Earth hours), but orbits leisurely around the Sun, some 1,427,000,000 kilometres away, just once every 29.46 years. Though more than 800 Earths might be packed inside it, its mass is only 95.14 times that of our world.

Voyager 1 was a billion miles from base and had been travelling for more than three years when it first began to transmit photographs of Saturn. Scientists at NASA stood in silent awe looking at the images of this gem of the heavens. 'It's like unlocking the secrets of ages,' said one. 'It really is almost like being there.'

And so it is, as the remarkable photographs on these pages show.

Naturally enough, it was Saturn's famous ring-system that first caught the scientists' eyes. Indisputably one of the wonders of the universe, the system is 169,000 miles in diameter, but less than ten miles thick. Indeed, when viewed side on, the rings seem almost invisible, and it is no wonder that they have been referred to as 'the thinnest things in existence'.

From Earth, the rings look solid enough and for generations astronomers believed that they consisted of millions of tiny moonlets, some just a few inches across, and all divided into three distinct rings. *Voyager's* very first pictures soon dispelled such long-held ideas.

The spacecraft's photographs revealed that there are, in fact, at least 300 ringlets consisting of solid particles of ice and rocks not merging gradually one into another as conventional theory would have us believe, but separated by complicated gravitational interactions between the ring material and six of Saturn's 15 (possibly 16) moons. Whether they were created when Saturn itself was forming billions of years ago and suddenly collapsed, throwing off

Opposite: *Breathtaking view of Saturn and its rings. Three images, taken through ultraviolet, violet and green filters, were combined to make this photograph.*

water vapour that froze into ice crystals, or are the remnants of a comet or moon that passed too close to the planet and was torn apart and swirled into orbit, remains a mystery.

For the first time, though, the penetrating gaze of *Voyager's* cameras enabled the scientists to clearly distinguish the various divisions of the rings. On the outer edge of the system is the E Ring, a broad, diffuse band of tiny particles, almost too faint to be noticeable except from close range and in which orbit the moons, Tethys and Enceladus. Below this, in the G Ring, which is similarly narrow and diffuse, circles the innermost moon, Mimas; while closer still to the planet, in the much narrower F Ring are several 'shepherd' moons which appear to be herding the entire main ring system and keeping within bounds any particles trying to escape. While studying this ring, *Voyager* came up with some astonishing data which seemed to defy the laws of gravity. For it appeared the band had a braided structure, like a three-stranded rope, with two of the strands intertwined and kinked. A theory has been advanced that the interaction of the tiny moons account for this peculiar phenomena.

There were, though, no surprises to be found when the spacecraft turned its sensitive equipment onto the A Ring, one of the three longest discernible divisions: a yellowish-tinted belt about 10,000 miles wide, in which lies Encke's Division, a very faint line about three-fifths of the way towards its outer rim. The A Ring is divided from its next partner by the famous Cassini's Division which can be seen from the Earth as an empty space, but which has now been shown to contain at least three dozen evenly spaced bands of ringlets.

In the 16,500 miles wide cream-coloured B Ring, however, *Voyager* came up with another puzzle for the NASA scientists to ponder over. For its cameras clearly picked out some curious spokes which radiated across the entire Ring. Once again, according to conventional space theory, this is not possible – but a theory has been put forward that the spokes might be made up of particles affected by electromagnetic forces, and it is these which cause them to form and then gradually fade away.

C Ring – or the Crêpe Ring as it is sometimes referred to – has always appeared relatively transparent, which explains why it was not discovered until as late as 1850. But *Voyager* went much further and showed it to be made up of dozens of dark ringlets, and at least one inexplicably non-circular ring. This whole ring is about 10,000 miles wide, and before the arrival of the NASA spacecraft, the remaining 9,000 miles gap to the outer reaches of Saturn's surface was thought to be empty. But *Voyager's* sensitive equipment and cameras confirmed what has now been designated the D Ring, a belt that clearly divides the ring system from the planet's atmosphere.

And what did the mission learn of Saturn itself, the second largest world in our solar system?

Putting ourselves in the position of the spacecraft, it becomes immediately evident from the rather leaden yellowish glow all around that we are dropping into a cold and silent world where night has been banished by the glow of the rings. The temperature is as low as −185°C at the cloud tops. However, beneath this bland appearance a haze may well be hiding features that are as colourful as those on Jupiter – though it is very hard to distinguish anything.

In making its fly-by at just over 30,000 miles from the edge of the rings, *Voyager* has been able to show that the planet has a number of white, oval spots on its surface and bands of lighter and darker clouds indicating equatorial and temperate belts. The sensors also picked up the movement of gigantic storms not unlike those experienced on earth – except, that is, the winds creating them were rushing along at speeds of up to 900 mph!

Saturn is believed to have a core of liquified rock under intense pressure with an ice sheet above this and an overlaying gas layer consisting mainly of hydrogen and helium. And in the thick, freezing, evil-smelling ammonia and methane clouds which swirl amidst all the planet's surface turbulence, temperatures may well fall to below −250°C. The chances of finding any form of life here, the scientists believe, are remote indeed.

Life may, though, just possibly exist on the largest of Saturn's 15 moons, Titan, and again *Voyager's* study added immeasurably to our knowledge of these icy, spiralling satellites which range from the biggest with a dimension of 3,000 miles to the smallest with less than 100 miles. The spacecraft's report on this cluster may be summarised as follows.

Closest to the planet lies a group of five moons designated by the year they were found (1980) and the numbers S 28, S 27, S 26, S 1 and S 3, while further out between Tethys and Dione is another probable moon noted as

Above: *A close-up view of Saturn's rings, colour-enhanced by computer.*

Opposite: *This montage of Saturn and its satellites combines individual photos taken by Voyagers 1 and 2. Mimas is outlined against Saturn, underneath the rings.*

1980 S 13. It was one of two such satellites seen for the very first time by *Voyager*.

The nearest named moon to Saturn is Janus, just 98,000 miles away, and for long considered the most elusive of all the satellites. Indeed, it was only discovered in 1966 by A. Dolfus at the Pic du Midi Observatory. With a diameter of 190 miles it orbits Saturn once every 17 minutes.

Mimas, at 113,300 miles from the planet, is a low density moon made of ice and frozen ammonia which is so cold that it has been described as rather like a huge snowball in space. But when *Voyager* began to transmit pictures back to Earth, what really surprised the NASA scientists was the sight of an enormous impact crater on the scored surface of the 350-mile diameter satellite. And in the centre of this 130 kilometre hole stood a central peak rising for some six kilometres from the floor.

202

One observer peered at this strange landmark and asked half in jest, 'Is that an engine?' Subsequent to this, the little world was nick-named by some of the NASA scientists 'The Death Star' because of its remarkable resemblance to the huge space machine in the movie, *Star Wars*. Look and judge for yourself!

It seems probable, though, that this curious formation on Mimas is the evidence of a collision between the moon and another satellite which very nearly blew it to smithereens.

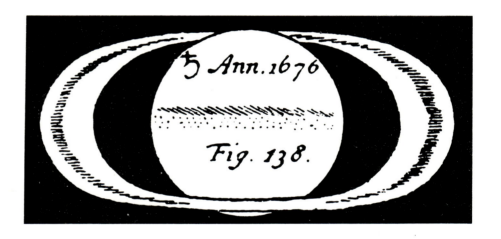

Drawing of Saturn by Giovanni Domenico Cassini, showing the division named after him, and dated 1676.

The neighbouring moon of Enceladus (not to be confused with the Mexican drink), is just over 35,000 miles away and slightly larger in size with a diameter of 450 miles. It is also a very low density, frozen world with a smooth, dazzling white surface which gives it the distinction of being the most reflective body in the solar system. It has been suggested that Enceladus keeps this shining appearance due to ice crystals seeping through the crust, or alternatively by having them scattered over the surface as a result of the constant impact of meteorites.

Tethys, almost half as big again with a diameter of 750 miles and located 183,00 miles from Saturn, is composed of over 80 per cent water and ice. It is covered with craters and a strange, branching trench mark 65 kilometres wide stretching from one end of the moon to the other. 'It's just like an old, chewed tennis ball,' one NASA scientist observed without much feeling.

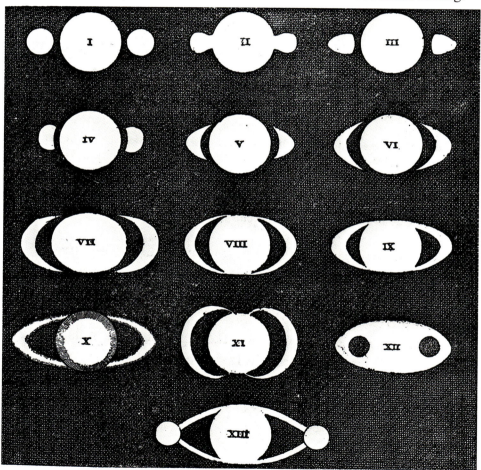

The various old ideas of Saturn – the first as three closely grouped planets, and on to the final realisation that it was circled by rings.

Next in line, Dione, is a more massive body, 900 miles in diameter and about as dense as our own Moon. Named after the mother of Aphrodite, it circles about 234,600 miles from Saturn, and has very worn features including innumerable fissures and craters. *Voyager's* photographs of strange, wispy markings on the surface offered strong evidence of gases having exploded out from the interior at some time in the past.

Rhea, another 93,000 miles further out, is also deeply cratered by bombardment and is criss-crossed by white trails indicating that water once burst from the interior. Being larger (800 miles in diameter) and brighter than all but one of the other moons – as well as having the greatest density of them all – Rhea is easily visible by telescope from Earth during its four-day orbit around the mother planet.

It was the next door moon of Titan which provided another moment of excitement for *Voyager* and those studying the data it transmitted. This, the largest of Saturn's moons, orbiting 759,500 miles away, proved a complete contrast to the other barren, icy satellites. For not only is it the largest moon in the solar system (as well as being bigger than the planet Mercury with a diameter of 3,000 miles) but on it the spacecraft registered an *atmosphere*, albeit 1.6 times denser than Earth's. The main constituent was analysed as nitrogen – making it the only place in the solar system apart from Earth to possess this gas. And because we know the oxygen in our atmosphere is largely the product of living things, there is now speculation that Titan might just be a primordial Earth!

Although Titan remained stubbornly shrouded in a reddish-brown haze as *Voyager* passed, the spacecraft was still able to detect in the atmosphere substances like acetylene, ethane and hydrogen cyanide (a critical building block for the more complex molecules of life) proving that some kind of synthesis of organic materials was occuring. According to scientists, a liquid nitrogen sea would provide an environment in which these materials could accumulate and interact. The potential for life does, therefore, seem to exist here. One school of thought has suggested that Titan's thick clouds might trap enough heat from the Sun to make life possible, while another has postulated that it could evolve in one of the immense oceans thought to exist beneath the satellite's frozen crust. The possibilities are obviously very intriguing.

As *Voyager* sped still further away from Saturn it next passed Hyperion, some 920,000 miles out, a tiny (200 miles in diameter) moon orbiting once in every 21 days. Sadly, though, it proved an undistinguished satellite and registered little of interest on either the spacecraft's cameras or instruments. A million miles on, bulky Iapetus put on more of a show for the terrestrial visitor. This 1,000 mile diameter ice-bound moon has the peculiar quality of one hemisphere being five times brighter than the other – and when *Voyager's* photographs were treated to enhancement back on Earth, the NASA scientists found themselves looking at something 'like a frizzled snowball in full flight'. The cause of this phenomenon could be an irregularity in the shape of the satellite or just one half of the body being more reflective than the other.

Far out into the yawning void into which *Voyager* then plunged lay Phoebe, 8,053,400 miles from Saturn and the last of its satellites. Laboriously circling the planet in a retrograde circuit, the 100-mile-diameter moon takes a mammoth 550 days to complete its circuit. Indeed, some scientists believe it may well not be a genuine satellite at all, but an asteroid captured by Saturn's gravitational pull. Whatever the case, it adds just a little more glitter to the remarkable 'Gem of the Heavens'.

URANUS & NEPTUNE
The Ends of the Galaxy

THE *Pioneer* and *Voyager* missions which blazed the trail of discovery as far as Jupiter and Saturn, are soon to be climaxed when *Voyager 2* flies on to the furthest points of the solar system and sends back information about the last two remaining planets, Uranus and Neptune, which it is scheduled to reach in 1986 and 1989 respectively. The results of this eyewitness tour so far have been remarkable, and there is no reason to suppose that what will follow will be any less spectacular. (*Voyager 1*, incidentally, is targetted to fly directly on towards deep space, of which more later.)

It seems to me, in fact, curiously appropriate that *Voyager* should be preparing for these rendezvous just as the first Britons go into space. For both worlds were actually first discovered by Englishmen! If fate can ever be said to play a hand in the affairs of men and nations, here is surely a most striking example of it!

Uranus, the first of the two planets on the far frontier of our part of the galaxy, was originally discovered just over two hundred years ago by the English astronomer, Sir William Herschel (1738–1822), and what he first saw so indistinctly through his home-made 6.2 inch reflector telescope is going to be brought into vivid, full-colour focus when *Voyager* and its highly sophisticated equipment arrives there in January 1986.

The planet, circling 1,775 million miles from the Sun, is the third largest body in the solar system. And although its distinctive blue-green disc has a

Opposite: *An artist's conception of the planet Uranus and its rings.*

209

diameter of 51,800 kilometres (four times that of Earth) and its reflective surface is not all that difficult to view with the naked eye, it wasn't discovered until the eighteenth century – and then Sir William at first thought it was a comet. Herschel was an instrument maker with a passion for astronomy and happened to be studying the heavens with his telescope from his private observatory in Bath when he came across *something* in the Gemini constellation.

His diary entry for a March evening in 1781 tells us what it was – or what he *thought* it was. 'There I saw a curious nebulous star,' he wrote, 'that being so much larger than the rest I suspected to be a comet.' Sir William was a conscientious observer, however, and continued his study. When he determined that the body had no tail and moved very slowly, he realised his first impression had been wrong. Instead he had found a new planet – the first man to have done so since classical times.

Herschel's discovery aroused enormous excitement, and initially it was to be named after his patron, King George III. But the French refused to accept this and for a time actually called the world after its discoverer. It was a German astronomer, Johann Bode, who suggested Uranus, taking the name from the Roman god who had been father to Saturn and grandfather of Jupiter.

The finding of this new world literally changed Herschel's life. He was knighted, appointed the King's Astronomer, and became the first president of the Royal Astronomical Society. Perhaps more importantly still, he founded 'Stellar Astronomy' (the study of stars), pushed back the known frontiers of the solar system far beyond their ancient limits, and made many important observations on the universe. He even made further discoveries in the vicinity of Uranus by discovering what proved to be two of the planet's five moons – Oberon and Titania.

Though a serious man by nature, and an immensely skilled instrument maker, Sir William was not above the odd flight of fancy, and later in his life quite seriously suggested that there were a race of people living in a protected region just beneath the fiery surface of the Sun!

We have, of course, learned a great deal more about Uranus since

Opposite: *Neptune and its largest moon, Triton, as visualized by an artist.*

Herschel's day – but much more remains to be discovered by *Voyager 2*. What we do already know is that it is, in effect, a giant gas world similar to Jupiter and Saturn. It has a dense hydrogen atmosphere which is many thousands of kilometres thick, covering a mantle of ammonia, methane and water over a small rocky core. The surface as such is actually the top of the freezing gas-cloud layer and the temperatures here range from −250°F to −350°F. The planet also has nine rings which are undetectable except with the most specialised equipment.

Perhaps, though, the most unusual thing about Uranus is that it orbits the sun, 2,870,000,000 kilometres way, *lying-down*! Its axial tilt relative to the plane of its orbit is 98 degrees, which means that first one rather flattened pole and then the other points directly at the Sun during the course of its enormously long (84 years) orbit. By contrast, it rotates swiftly on its axis, once in about 15.6 hours. This results in the most bizarre 'seasons' on the planet. For periods of 21 years much of the northern hemisphere, and then much of the southern hemisphere, will be plunged into the most intense darkness, with a corresponding midnight sun in the opposite half of the world. For the remaining 42 years night and day conditions will be as normal.

Intensive astronomical study of Uranus has shown that it has a number of greyish belts on its surface as well as some dark spots – although the most distinctive features are a white equatorial zone and rather dusky-looking poles. The planet's low density means that although it is big enough to

An observation of Uranus and the Satellites of Jupiter made on 5 June 1872 – from Camille Flammarion's book, Les Terres du Ciel.

212

contain 64 Earths, it would only weigh as much as 15 of them. And any human being on the 'surface' would actually feel slightly lighter than on Earth.

There are suggestions that there may well be disturbances taking place within the planet, but only the closer look that *Voyager* can provide will answer this question. The spacecraft will similarly be able to give us valuable data on the five moons which all revolve close to the plane of Uranus' equator at distances between 364,000 miles and 76,000 miles.

The furthest out of these satellites are Oberon and Titania which Sir William Herschel discovered. The first of these has a diameter of 920 kilometres and takes 13 days to orbit the planet, while the larger Titania at 1,100 kilometres makes its circuit in 8 days. The next pair of moons, Ariel and Umbriel, are so close to the mother planet that they are lost in its glare for much of the time. It took another Englishman, William Lassell, at Liverpool, to pinpoint their existence in 1851. Umbriel, 640 kilometres in diameter, orbits Uranus in just over four days, while slightly larger Ariel, 800 kilometres in diameter, takes 2.5 days. (Interestingly, all four of these moons were named by Sir John Herschel, son of the planet's discoverer.)

The fifth and last of the satellites defied all efforts to trace it until less than fifty years ago. Miranda is so tiny and so close to Uranus that it cannot be seen visually, and it took dedicated work by the American Gerald Kuiper at the McDonald Observatory in Texas to finally prove its existence with a series of photographs he took in 1948 with the aid of his massive 82 inch telescope.

It is perhaps not surprising to learn that the discovery of Uranus led very quickly to the finding of Neptune, the last of the giant Jovian planets, almost a thousand million miles on from its neighbour. This will be *Voyager 2's* final point of call in the solar system in September 1989.

As I mentioned earlier, another Englishman was the first to suspect the existence of the planet. He was an astronomer named John Couch Adams (1819-1892) who came up with the idea in 1841. After several years of star gazing, he confided in his Diary for the July of that year:

'Formed a design, in the beginning of this week, of investigating, as soon as possible after taking my degree, the irregularities in the motion of Uranus which are yet unaccounted for; in order to find whether they may be attributed to the action of an undiscovered planet beyond; and if possible thence to determine the elements of its orbit, etc., approximately, which would probably lead to its discovery . . .'

Unfortunately young Adams was frustrated in his endeavours to mount a serious search for the planet by the demands of his studies at Cambridge and by the general scepticism which existed towards such an idea in English astronomical circles. Nonetheless he deposited a paper with the Astronomer Royal at Greenwich in October 1845 outlining his belief. A year later, after some intermediary work by a French astronomer, Urbain Leverrier, who also had the same suspicions, two Germans established the existence of Neptune in October 1846. The men, Johann Galle and Heinrich D'Arrest, who worked at the recently opened Berlin Observatory, generously paid tribute to Adams and his work in enabling them to confirm the finding of the second new world in modern times. Adams himself later became Professor of Astronomy at Cambridge and made some more important space discoveries – albeit rather closer to home – about the Moon.

Neptune has proved to be rather similar to Uranus in various ways, and some scientists have even called them twin worlds. An almost identical greenish hue, it has a similar diameter at 48,600 kilometres, and a composition consisting of a dense hydrogen atmosphere (3,000 kilometres deep) overlaying a mantle of methane, ammonia and water (10,000 kilometres) and a large rocky core 20,000 kilometres thick.

This incredibly remote world is 4,504,000,000 kilometres from the Sun which it orbits once every 165 years. A single rotation, however, takes a remarkably short 17.9 hours, and like Uranus its orbit is tilted – though to a rather lesser extent of 29 degrees – so that the extraordinary seasonal conditions are not repeated here. Neptune is the densest of the four Jovian worlds, and with a higher surface gravity than its 'twin' it would add as much as half his weight again to any humaan visitor. Curiously, the cloud top temperature estimated at −205°C is higher than might be expected for a planet this far from the sun, and some scientists have suggested that the methane may well create a 'greenhouse' atmosphere.

There is also a growing conviction that Neptune may well be like its three

Opposite: *Artist's conception of Pluto and its moon Charon. The white spot in the distance is the Sun.*

fellows and have a ring system, and *Voyager's* arrival to confirm or deny this is eagerly awaited. Of the planet's features, a bright equatorial zone has been observed, while the poles seem somewhat darker with hints of a belt structure. Because of the intense cold, it is believed that this is a very quiet world with little or no surface activity; but again *Voyager* may surprise us.

The first of Neptune's moons, Triton, was spotted by William Lassell who did such useful work on Uranus' satellites. He announced his discovery just a month after the two Germans confirmed the existence of the planet itself. Triton is said to share a number of similarities with Saturn's intriguing moon, Titan, although it is a little smaller in diameter at 3,700 kilometres. The

The size of Neptune as compared with Earth – also by Camille Flammarion.

strongest similarity is in the atmospheric mantle believed to consist of methane. The little moon circles the mother planet at a distance of 355,000 kilometres once every five days – but quite inexplicably in a strongly elliptical and clockwise direction! It is the only satellite of its size we know to do this.

The other moon, Nereid, is very minute and invisible to all but the very strongest astronomical equipment. It was again the redoubtable Gerald Kuiper in Texas who spotted it for the first time on photograph in 1949. Just under 500 kilometres in diameter, Nereid orbits Neptune (5,562,000 kilometres away) once every 360 days. This orbit is a most unusual one, however, more like that of a comet that a satellite – for at its closest it comes to within a million miles of the mother planet, while at its farthest point it is well over six million miles away! Once again we may hope for some explanation for this from *Voyager*.

And so to the very furthermost reaches of our solar system where at a point 5,900,000,000 kilometres from the Sun circles the lonely, isolated, enigmatic little world known as Pluto. Considered to be a genuine planet by some authorities, there are others who are equally convinced that it is no more than an escaped moon that once belonged to Neptune. Support is leant to this theory by the fact that its known constituents are methane and water mixed with rocks.

Whichever may eventually prove to be the case, Pluto is murky yellow in colour and at approximately 3,000 kilometres in diameter, smaller even in planetary terms than Mercury. It takes a mammoth 248 years to revolve in its elliptical orbit around the Sun, while it rotates on its own axis in 6.4 Earth days.

Once again, this world of eternal half-light was not observed until this century by Clyde Tombaugh at the Lowell Observatory in Flagstaff, Arizona, in 1930. He was also able to show that this 'snowball-like' body also spends a twenty year period during its revolution passing *inside* Neptune's orbit. A satellite, named Charon, has also been spotted, but as *Voyager 2* goes nowhere near this pair, we may have to wait generations yet before receiving an eyewitness report from the very end of our galaxy of worlds.

Instead, *Voyagers 1* and *2* will leave our solar system in about 1990, and plunge into the vastness of deep space. They will follow *Pioneers 10* and *11* as the third and fourth human artifacts to escape entirely from the gravitational pull of the Sun. The *Pioneers* carried small metal plaques identifying their time and place of origin for the benefit of any other spacefarers that might find them in the distant future (see illustration). Also included was data on the human species and our biological form.

Pioneer achieved the kinetic energy needed to carry it out of the solar system when it flew by Jupiter, and it is hoped by NASA that between one

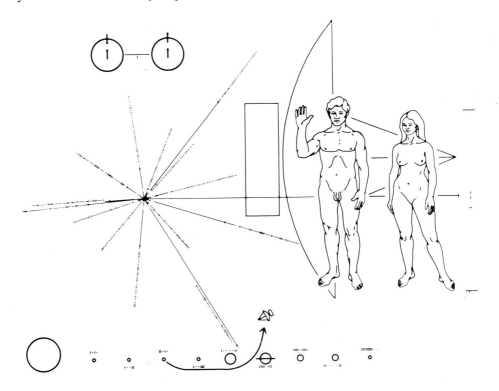

Above: *The* Pioneer *Spacecraft carries this plaque on its journey beyond our solar system, bearing data that tells where and when the human species lives and containing details of our biological form.*

Opposite: Pioneer *leaving the solar system. The planets are shown in their relative positions (but at exaggerated scales), as they were on 13 June 1983, the day* Pioneer 10 *passed outside the orbit of Neptune (Pluto, due to its elliptical orbit, was inside Neptune's orbit) and thus left the solar system.*

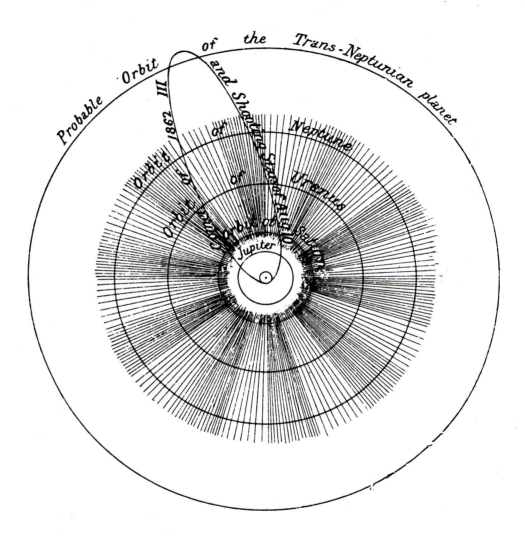

A nineteenth century speculation on the probable existence of Pluto.

and ten billion years from now it may pass through a remote planetary system where a world may have evolved a form of life capable of detecting and inspecting the probe.

The *Voyagers* carry an even more sophisticated message that the scientists want to reach across the eons to communicate information about us. These are 'time capsules' – 12-inch gold plated copper phonograph records containing sounds and images selected to portray the diversity of life and culture on Earth. The sounds include greetings in 60 languages, music from different nations and eras, and the noise of wind, surf, animals and many other earth phenomena. The 115 images range across views of human activities to geological formations and man-made constructions as well as our alphabet and numbering system.

The contents of these 'time capsules' were selected by a group of scientists headed by Dr Carl Sagan, and after the two *Voyagers* had departed, he commented, 'The spacecraft will be encountered and the records played *only* if there are advanced spacefaring civilisations in interstellar space. But the launching of these bottles into the cosmic ocean says something very hopeful about life on this planet.'

And as we mark the journey of the first Briton into only the very *closest* reaches of space, I should like to say 'Amen' to that and add the singularly appropriate words the great British poet, T. S. Eliot, wrote just a few years ago:

> *Not fare well*
> *But fare forward, Voyagers.*